T0123102

THE **EAR** BOOK

A Johns Hopkins Press Health Book

Thomas J. Balkany, MD, FACS, FAAP, is Hotchkiss Professor and Chairman Emeritus, Department of Otolaryngology, and formerly Professor of Neurological Surgery and Pediatrics, University of Miami Miller School of Medicine. Dr. Balkany is board certified by the American Board of Otolaryngology in Otolaryngology as well as in Neurotology. He has published four books and over two hundred fifty scientific papers, and he holds fourteen US and international patents. His research has been supported by the National Institutes of Health, and he was a panelist of the US Food and Drug Administration. Dr. Balkany has served on the editorial boards of eleven indexed peer-reviewed journals and was an associate editor of *Audiology and Neurotology*. He has also acted as a senior examiner of the American Board of Otolaryngology and is currently Director, Institute for Cochlear Implant Training, where he teaches advanced surgical techniques in cochlear implant surgery.

Kevin D. Brown, MD, PhD, is Associate Professor of Otolaryngology and Neurosurgery at the University of North Carolina School of Medicine. He is also the medical director of the Children's Cochlear Implant Center at the University of North Carolina, the largest pediatric implant program in the United States. He is Board Certified in Otolaryngology and Neurotology. He has been published in numerous clinical and scientific papers and has written chapters for several textbooks. He serves as an examiner for the American Board of Otolaryngology and is on the Editorial Board of the major journal for clinical ear research, *Otology and Neurotology*.

THE **EAR** BOOK

A Complete Guide to Ear Disorders and Health

Thomas J. Balkany, MD, FACS, FAAP

Kevin D. Brown, MD, PhD

JOHNS HOPKINS UNIVERSITY PRESS

Baltimore

Note to the Reader: This book is not meant to substitute for medical care and treatment should not be based solely on its contents. Instead, treatment must be developed in a dialogue between the individual and his or her physician. Our book has been written to help with that dialogue.

Drug dosage: The author and publisher have made reasonable efforts to determine that the selection of drugs discussed in this text conform to the practices of the general medical community. The medications described do not necessarily have specific approval by the US Food and Drug Administration for use in the diseases for which they are recommended. In view of ongoing research, changes in governmental regulation, and the constant flow of information relating to drug therapy and drug reactions, the reader is urged to check the package insert of each drug for any change in indications and dosage and for warnings and precautions. This is particularly important when the recommended agent is a new and/or infrequently used drug.

© 2017 Thomas J. Balkany and Kevin D. Brown
All rights reserved. Published 2017
Printed in the United States of America on acid-free paper
9 8 7 6 5 4 3 2 1

Johns Hopkins University Press
2715 North Charles Street
Baltimore, Maryland 21218-4363
www.press.jhu.edu

Library of Congress Cataloging-in-Publication Data
Names: Balkany, Thomas J. | Brown, Kevin D. (Kevin David)
Title: The ear book : a complete guide to ear disorders and health /
    Thomas J. Balkany, MD, FACS, FAAP, and Kevin D. Brown, MD, PhD.
Other titles: Complete guide to ear disorders and health
Description: Baltimore : Johns Hopkins University Press, 2017. | Series: A
    Johns Hopkins Press health book | Includes bibliographical references and
    index.
Identifiers: LCCN 2016040194| ISBN 9781421422848 (hardcover : alk. paper) |
    ISBN 9781421422855 (pbk. : alk. paper) | ISBN 9781421422862 (electronic) |
    ISBN 1421422840 (hardcover : alk. paper) | ISBN 1421422859 (pbk. : alk.
    paper) | ISBN 1421422867 (electronic)
Subjects: LCSH: Otology—Popular works. | Ear—Diseases—Popular works. |
    Hearing disorders—Diagnosis—Popular works. | Hearing
    disorders—Treatment—Popular works.
Classification: LCC RF123 .B245 2017 | DDC 617.8—dc23
    LC record available at https://lccn.loc.gov/2016040194

A catalog record for this book is available from the British Library.

Figures 1.1, 1.2, 1.3, 4.1, 5.1, 8.1, 9.1, 9.2, 10.1, 12.1, 13.1, 13.2, 14.1, 15.1, 18.1, 19.1, 20.1, and 29.1 are by Jane Whitney.

Special discounts are available for bulk purchases of this book. For more information, please contact Special Sales at 410-516-6936 or specialsales@press.jhu.edu.

Johns Hopkins University Press uses environmentally friendly book materials, including recycled text paper that is composed of at least 30 percent post-consumer waste, whenever possible.

To my family—Diane, Jordan, and Sarah Balkany
                                    —TJB

To my wife, Nicole, and our three sons, Luke,
Gabriel, and Ethan
                                    —KDB

Contents

IV | Disorders of the Middle Ear

V | Disorders of the Inner Ear

VI | Other Things You Should Know

Preface

The importance of any organ system of the human body is related to how often it must be addressed by health care professionals. The most common illness requiring a prescription from the pediatrician, the most common operation performed on children, and the most common disability affecting adults in the United States all involve the ear. *The Ear Book* is a comprehensive explanation of the causes, diagnosis, and treatment of these frequently occurring ear disorders. It is written in plain language so it can be useful to all those managing this organ system, including general health care providers and the book's primary audience—the lay public.

Symptoms caused by conditions affecting the ear such as hearing loss, ringing in the ears, and dizziness can be disabling. Explanations of the cause of the problem and the need and value of testing, as well as treatment options, can be difficult to follow in the physician's office. Even common problems like ear wax, swimmer's ear, ear infections, and perforations of the eardrum can be painful and confusing.

Rapid advances in medicine make it difficult to keep up, even for specialists in the field. Exaggerated claims of miraculous cures found in the media may be difficult for patients and providers to separate from established treatments founded in good clinical science. We describe recent innovations in simple terms and sift fact from fiction to help patients and their caregivers stay abreast of modern approaches to the ear.

What You Will Find in This Book

The ear is one of the most complex organs of the body, providing the sense of hearing and balance. Part 1 begins with clear explanations of how the ear works. After a review of the meaning of different ear symptoms, we dispel some popular myths about the ear.

In part 2, we review the most prevalent ear disorders—otitis media, hearing loss, dizziness and vertigo, and tinnitus—focusing on finding the right diagnoses and getting the best treatment.

Part 3 looks specifically at problems affecting the outer ear—swimmer's ear, obstructions, malformations and growths, cancer, and trauma. The middle ear is the focus of part 4, with a closer look at perforations of the eardrum, otosclerosis, and mastoiditis. Part 5 looks into the inner ear. Here we discuss Ménière's disease and noise-induced hearing loss, two common problems. We also describe the symptoms of sudden deafness and autoimmune diseases as well as the consequences of using some life-saving drugs. We close part 5 with a discussion of tumors.

In the final section we describe different types of hearing and balance tests. Knowing what to expect in advance will help you get the most out of your visit to the doctor. In this section we also address the problems of adjusting to different air pressures (when scuba diving or flying in a plane). Techniques can be learned to lessen the effects of pressure changes. The last chapter looks to the future and the progress that has been made with gene therapy and stem cell research.

At the end of the book you will find a glossary of terms, appendixes providing a list of ear medications and a roster of resources, references for further reading, and an alphabetical index to facilitate easy access to specific topics.

The Ear Book provides up-to-date, reliable information. It follows a conservative approach to surgery. We hope that it will help you in working closely with your primary physician or ear specialist.

I | Understanding Your Ears

1 | How the Ear Works

A few minutes ago every tree was excited, bowing
to the roaring storm, waving, swirling, tossing their
branches in glorious enthusiasm like worship. But
though to the outer ear these trees are now silent,
their songs never cease. —JOHN MUIR

The ear is one of the most complex and compact parts of the body. This chapter is a guide to the normal structure and function of this organ that is responsible for hearing and balance. In subsequent chapters we'll explore disorders of the ear.

Our ears have three basic parts: outer ear, middle ear, and inner ear.

- The *outer ear* funnels sound waves to the middle ear.
- The *middle ear* amplifies the sound waves and delivers them to the inner ear.
- The *inner ear* converts the sound waves to nerve signals and sends them on to the brain where they are "heard" (see figure 1.1). The inner ear is also responsible for balance. It detects head motion and gravity and sends signals to the brain to provide our sense of equilibrium or balance.

Outer Ear

Pinna (Auricle)

The outer ear consists of the visible part of the ear (*pinna*) plus the ear canal—the place we *don't* put cotton swabs. Sound waves are funneled by the pinna into the ear canal, but sounds from behind are partially blocked by the pinna, allowing a listener to focus on sounds in front. The pinna is formed by cartilage covered with skin. Small muscles are attached to the back of the ear, allowing some people to wiggle theirs (and making it possible for dogs and cats to turn theirs toward sound). Disorders of the pinna, from cauliflower ear to skin cancer, are discussed in part 3.

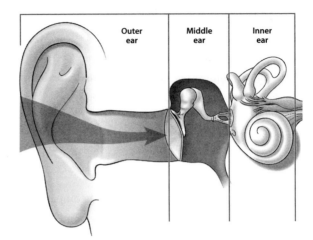

Figure 1.1
The three parts of the ear are the outer ear, middle ear, and inner ear.

Within the figure:
- Outer ear
- Middle ear
- Inner ear

External Ear Canal

The *ear canal* carries sound waves to the eardrum. There are glands in the skin covering the cartilage of the outer third of the canal. These glands produce wax that coats and protects the skin lining the canal. Skin over bone makes up the inner two-thirds of the canal, which has no glands. The external ear canal is about an inch long and a third of an inch wide, which is the perfect diameter for resonance of the human voice frequencies. Since the outer ear canal is crooked, it helps prevent swabs, pencils, and other hard objects from reaching the eardrum. The eardrum seals off the inner end of the ear canal and marks the beginning of the middle ear.

Middle Ear

Eardrum

The *eardrum* forms a watertight seal between the ear canal and middle ear and picks up vibrations of sound waves in the ear canal. It is about 9 mm (1/3 inch) in diameter and about 0.075 mm (3/1000 inch) thick. On the outside it is covered with skin and on the inside with mucous membrane, like the nose and throat. The middle layer of the eardrum is strengthened by fibrous tissue and contains many tiny blood vessels.

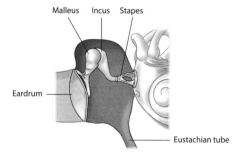

Figure 1.2
Three bones of the middle ear are the *malleus* (hammer), *incus* (anvil), and the *stapes* (stirrup). The malleus is attached to the eardrum. The incus is second, and the stapes attaches to the inner ear (cochlea).

Ossicles (Three Bones of Hearing)

On its inside surface, the eardrum is attached to the first bone of hearing, called the *malleus* or hammer. Sound waves in the ear canal vibrate the eardrum, which in turn vibrates the malleus and two other bones of hearing, the *incus* (anvil) and the *stapes* (stirrup). These middle ear bones are the three smallest in the body and have the two smallest muscles in the body attached to them. Those muscles can dampen (reduce) vibrations in certain loud situations. Together the eardrum and middle ear bones amplify sound about twenty-six times before delivering it to the inner ear.

Eustachian Tube

The middle ear is lined with mucous membrane and normally filled with air. That air space is connected with the nose and throat by the *Eustachian tube*. The tube is normally closed to prevent food and saliva from running into the middle ear. But when we swallow, yawn, and chew, tiny muscles open the Eustachian tube automatically.

The two main functions of the Eustachian tube are to allow mucous to drain out of the middle ear and air to move into and out of the middle ear. Passage of air equalizes middle ear pressure with the air pressure of the outside world. This is important during airplane travel and scuba diving.

Middle Ear Disorders

The most common problems of the middle ear are infection and fluid buildup. These occur frequently in children partly because their Eustachian tubes do not mature until pre-adolescence. Perforations, bony fusion or dislocation of the bones, and tumors can also affect the middle ear (for more on these disorders see part 4).

Inner Ear

The inner ear has two parts: the *cochlea* (for hearing) and the *balance system*. Both contain sensory receptors called *hair cells* that are suspended in inner ear fluid. The hair cells of the cochlea distinguish sound vibrations in the fluid and change them into electrical nerve signals that transmit to the hearing areas of the brain. The hair cells of the balance system perceive movement and gravity and change them to electrical nerve signals that travel to the balance sectors of the brain.

Cochlea

The cochlea is less than 1 cm (about 0.4 inches) across. It looks like a nautilus shell, with three fluid-containing compartments arranged in a spiral. Over fifteen thousand hair cells in each cochlea are moved by the vibrations of inner ear fluid. The motion causes inner hair cells to produce electrical signals that transmit to the brain and are perceived as hearing.

Balance System

The shape of the balance system—also called the vestibular labyrinth—is complex. Each inner ear has three *semicircular canals*, a *utricle*,

Figure 1.3
The inner ear is snail shaped and contains fluid, sensory cells, and nerves. The cochlea receives vibrations from the middle ear and the balance canals are stimulated by gravity and motion. These stimuli are changed by the inner ear balance system (transduced) into nerve signals and sent to the brain.

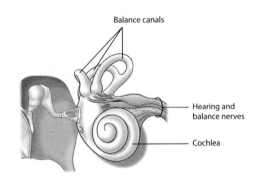

Balance canals

Hearing and balance nerves

Cochlea

Table 1.1
Summary of the Parts of the Ear and Their Function

Name	Description	Function
Outer Ear		
Pinna	Visible part of the ear	Funnels sound into ear canal
External canal	Leads to eardrum	Enhances speech frequencies
		Protects eardrum
Middle Ear		
Eardrum	End of ear canal	Amplifies, transfers sound to ossicles
		Creates air-containing space
Ossicles	Middle ear bones	Amplifies, transfers sound to cochlea
Inner Ear		
Cochlea	Fluid and hair cells	Converts vibrations to nerve signals
Balance system	Fluid, hair cells, crystals	Converts motion/gravity to signals

and *saccule*. The three semicircular canals perceive head movement, especially turning motion. The utricle and saccule respond to the pull of gravity rather than motion. All of these structures have tiny crystals that provide mass (almost the same as weight) to the balance system hair cells. When the head is moved into a new position, gravity pulls on that mass and bends the tips of the hair cells. Bending a balance system hair cell causes it to send a nerve signal to the brain that indicates head movement or position.

2 | **Ear Symptoms** | What Do They Mean?

■ A middle-aged man experiences intense vertigo (a spinning sensation) and nausea a few days after developing an upper respiratory infection. His hearing from the left ear has deteriorated, and he can no longer understand words on the telephone. There is also an intense ringing in his left ear, and it feels as if the ear is underwater.

This patient's history—the combination of nausea and vertigo, hearing loss, and tinnitus—tells the doctor that the man probably has a disorder of both the hearing and balance parts of the inner ear. The doctor will use this information to focus the medical examination and testing that will help lead to a diagnosis. The patient's history of symptoms is often the most important part of the medical evaluation. ■

Ear disorders can cause various symptoms. Each symptom alone can suggest certain problems, but when the indicators are looked at in combination, a clearer picture emerges. Understanding the symptoms is the key to knowing what is triggering ear disorders. In this chapter we discuss some of the more common ear symptoms and their causes.

Hearing Loss

Hearing loss is a reduction in the perception of sound by the brain. This loss can be caused by problems of the outer, middle, or inner ear, the nerve of hearing, or the brain itself. The precise cause is identified by conducting a review of the patient's medical history, a physical examination, and testing.

Sound can be blocked in the ear canal or middle ear by wax, fluid, infection, or middle ear bone disorders.

When the cochlea (inner ear) is not working, sound waves are not converted (transduced) into nerve signals that can stimulate the brain. Some explanations for cochlear problems include genetic abnormalities, noise exposure, certain antibiotics and anti-cancer drugs, and the process of aging.

The nerve of hearing, which connects the cochlea to the brain, can be damaged by tumors and head trauma. And the brain itself may not function properly as a result of infection, stroke, degeneration, or trauma.

Tinnitus

Tinnitus is any type of sound heard by a person that does not have a source in the outside world. When tinnitus is caused by disorders of the cochlea, auditory nerve, or brain it may be perceived as ringing, roaring, white noise, or even music.

Sometimes tinnitus sounds like the pulse or heartbeat. This kind of tinnitus is the result of turbulent blood flow in the arteries or veins inside the skull and can often be heard by the ear specialist as well as the patient. Pulsatile tinnitus and turbulent blood flow can be caused by high blood pressure, abnormal blood vessels, aneurysms, and tumors. It may also be a byproduct of *benign intracranial hypertension* (mild hydrocephalus) that is associated with obesity.

Clicking or typewriter-like tinnitus is caused by contraction of muscles near the ear. It may be the result of head trauma, stress, stroke, and rare tumors. Many people describe this kind of tinnitus as sounding like an old-fashioned typewriter.

Vertigo

Vertigo is a feeling of motion when you are holding still. Vertigo is often compared to imbalance and lightheadedness, but there are different root causes. For practical purposes it is generally true that the balance system of the inner ear causes the symptoms of vertigo, while imbalance and lightheadedness are likely to be the result of factors such as low blood sugar, thyroid disorders, circulation issues, drugs, or stroke. (See chapter 10 for more detail.)

The key to identifying the cause of vertigo is how long each episode lasts. Short, five- to ten-second, bursts of vertigo associated with head motion may be a result of *benign positional vertigo*, caused by dislocation of calcium crystals in the inner ear. Long-lasting vertigo, more than twenty-four to forty-eight hours, that improves slowly over a one- or two-week period is typically caused by vestibular neuritis. Intense, episodic vertigo lasting two minutes to a few hours with associated hearing loss, tinnitus, and ear pressure typically suggests Ménière's disease (see chapter 21).

Fullness

A common cause of a feeling of fullness or blockage of the ear is wax obstruction of the ear canal or fluid behind the eardrum. However, it can also come from abnormal fluid pressure in the inner ear, in which case a visual examination of the ear appears normal. A pressure sensation can also be referred from other locations.

Ear Pain

Ear pain typically is a sign of inflammation and/or infection. If the ear canal is draining, the fluid usually indicates an infection of the ear canal skin. If the ear canal is normal but yellowish discoloration is seen behind a red eardrum, it suggests that the middle ear is infected.

Ear pain may come from the ear itself or surrounding structures that have the same sensory nerve supply as the ear. Common examples of non-ear problems that feel like ear pain include disorders of the temporomandibular joint (TMJ), the throat, and the voice box. When non-ear disorders cause ear pain, the pain is called *referred pain*—that is, the source of the pain is elsewhere in the body but it is felt in the ear.

Ear Drainage

Infection of the ear canal or the middle ear usually results in fluid draining from the ear. It is typically yellow to green and smells bad. Your doctor will need to carefully clean and inspect the ear to determine the source. Cleaning the ear canal is also the first step in treatment because it allows antibiotic drops to work and reduces the infection.

Pus from a middle ear infection (called otitis media) may make its way to the ear canal if a perforation of the eardrum occurs. Such a perforation may be painful at first, but quickly becomes less painful—unlike the persistent pain caused by an ear canal infection (swimmer's ear). Inflammatory conditions such as eczema of the ear canal skin generally have a clear to light yellow drainage and produce more itching than pain. To complicate matters, inflammatory conditions like eczema can lead to recurring infections of the ear canal.

Hearing Your Own Voice

Autophonia, the sense of loudness or echoing of your own voice, may occur when the ear canal is occluded (by wax) or when fluid builds up behind the eardrum. To experience this sensation, place your palm firmly over the outer ear opening and speak a sentence. Your voice will sound louder in the blocked ear. This is called the occlusion effect. Other disorders—such as disruption of the middle ear bones or a too-wide-open Eustachian tube—may also cause autophonia.

Hearing Your Breathing

The Eustachian tube connects the ear to the upper airway, which continues all the way to the lungs (see chapter 1). It is normally a closed structure except when a person is swallowing or yawning. But with rapid weight loss, sometimes the Eustachian tube stays open. The sounds of breathing can then pass into the ear. This can be treated effectively with topical medication or, in severe cases, by partially blocking the tube.

3 | Common Myths about the Ear

■ Just like with headaches, there is tinnitus and there is TINNITUS! And like so many others, my doctor told me that the noise in my head was common and I should learn to live with it because nothing could be done. But he wasn't awake most of the night, fatigued, stressed-out, irritable, and depressed. It got worse and worse until I saw an ear specialist who did a complete evaluation. I felt better just knowing that there was no dangerous underlying cause. Just avoiding caffeine and aspirin reduced the severity by half. Working out each day and avoiding naps helped me sleep at night and gradually I began to feel healthy again. I learned about sound therapy for tinnitus, but I was already in good shape. ■

Sometimes misinformation is worse than no information at all, especially if it leads to misunderstanding and poor medical care. The following are some common myths that may lead to injury, delayed treatment, and unnecessary expense.

MYTH *Oral antibiotics are necessary to treat outer ear infections such as swimmer's ear.*

FACT Not true. Most outer ear infections, also called external otitis (see chapter 12), occur on the surface of the skin of the ear canal. Because the blood supply is poor there, oral antibiotics may not reach the infection in high enough concentrations to be effective. Antibiotic ear drops are a better choice because they contain high concentrations of powerful antibiotics that are applied directly to the infected area. Ear drops are also slightly acidic, reducing the growth of bacteria. Oral or injectable antibiotics may become necessary in people with low resistance to infection (people who have diabetes, transplant or cancer patients, and people who have HIV or AIDS) or when the bacteria begin to penetrate into the deeper layers of skin, cartilage, or bone.

MYTH *Ear wax is just dirt.*

FACT Not true. Ear wax, also called *cerumen*, is an important protective substance produced by glands in the skin of the ear canal along with skin cells that are shed daily, just like skin cells everywhere in the body. Cerumen exits the ear canal on its own during showering, swimming, and normal everyday activities and carries with it other matter that may have entered the canal. Cerumen's main functions are to protect the delicate skin of the ear canal from damage by moisture and to fight outer ear infections such as swimmer's ear.

MYTH *Cotton swabs are necessary to clean the ear canal.*

FACT Not true. The ear cleans itself in most cases. Because cotton swabs are about the same diameter as the ear canal, they often act as a plunger, pushing wax back into the ear canal and against the eardrum. Because they are long and stiff, swabs cause injuries to thousands of people every year. Often the skin of the ear canal is damaged by the swab, leading to infections. Sometimes the swab is poked through the eardrum, causing hearing loss, tinnitus, and intense pain. Severe injuries often occur when someone gets bumped on the elbow while using the swab at the bathroom sink; such injuries may require surgery to repair.

MYTH *Ear candling removes wax from the ear canal.*

FACT Not true. This multimillion-dollar industry is a magic act. Ear candles are hollow and are touted to create a vacuum when the wick is lit and the base is inserted into the ear opening. The vacuum is said to suck out ear wax, and the wax that accumulates in the base of the candle is supposed to be the proof. But careful studies have shown that the wax accumulation in the hollow candle is actually melted from the candle itself and has no cerumen in it. The candles do not create any vacuum and many people have been burned when the candle's wax drips down into the ear canal.

MYTH *Ear pain means ear infection.*

FACT Not necessarily true. While ear infections commonly cause ear pain, there are also other important reasons for pain. *Temporomandibular joint pain* (TMJ) is the most common cause of ear pain that does not come from the ear. TMJ pain is from inflammation or trauma to the joint that connects the jaw to the skull. This joint has pain fibers from the same nerve that serves the ear, so TMJ pain is often mistaken for ear pain. Disorders of the teeth, throat, and voice box may also cause pain that feels like it is coming from the ear, also because they share a common sensory nerve. This is technically called referred pain.

MYTH *Children with ear tubes must not go swimming.*

FACT True in some cases. While this is true with some types of ear tubes, it is safe for most children with tubes to swim in private, chlorinated, filtered pools. Ear plugs or head bands are not required by most physicians. But there are some limitations. The child's head should not be submerged more than 6 inches. And the ears should not be submerged in bathwater, lakes, rivers, kiddie pools, hot tubs, or at the beach. Your ear doctor will advise you on a case-by-case basis.

MYTH *Parents or the pediatrician will know if a child has hearing loss by two or three months of age.*

FACT Not true. Before the era of universal newborn hearing screening, the average age at which children were identified with severe hearing loss was two to three years. Children with mild hearing loss were usually not discovered until age four. Delay in diagnosis can hinder language development, and the results can be permanent. Today, hearing loss is usually diagnosed before twelve months of age. According to the Centers for Disease Control and Prevention, in 2011, 98 percent of American newborns were screened for hearing loss.

MYTH *Loud sound is not dangerous if it does not hurt.*

FACT Not true. Sound will damage hearing at 85 decibels (moderately loud machinery for eight hours) but pain may not be noted until levels of 110 decibels or more (jet plane taking off at 150 feet).

MYTH *Health food supplements can improve hearing and reduce tinnitus.*

FACT Not proven. Many supplements such as bioflavonoids, zinc, ginkgo biloba, and other antioxidants have been promoted as over-the-counter remedies for decades. Many studies have failed to show that these treatments work better than a placebo. However, it has been shown that any treatment for tinnitus that promotes a sense of well-being or reduces anxiety will also lessen the perception of tinnitus. The placebo effect is around 30 percent for tinnitus, so most supplements have helped someone.

MYTH *There is no treatment for tinnitus.*

FACT Not true. Tinnitus is a symptom, not a disease, and may be caused by several different disorders. Although there is no known cure for most tinnitus, it can be effectively treated (just as high blood pressure and diabetes can be treated if not cured). Most people can be helped, especially with early diagnosis. Most tinnitus is responsive to cognitive behavioral therapy, use of hearing aids, or other sound-based treatment. In rare situations, tinnitus may be caused by a dangerous underlying condition (such as a tumor or aneurysm) and require medical or surgical treatment. Individuals with pulsating tinnitus or tinnitus that is louder in one ear than the other should consult the doctor immediately since these are red flags.

MYTH *There is no treatment for nerve deafness.*

FACT Not true. Both hearing aids and cochlear implants can be very helpful for people with nerve deafness. Hearing aids are useful for mild-moderate-severe hearing loss and cochlear implants are safe and effective treatment for moderate-severe-profound hearing loss (see chapter 5).

MYTH *Hearing aids weaken the ear.*

FACT Not true. Hearing aids do not act like crutches. This is because the ear is not a muscle. It does not get weaker or atrophy when you use hearing aids.

MYTH *It is necessary to spend $4,000 to $6,000 to buy a good pair of hearing aids.*

FACT Not true. It is a fact that hearing aids are expensive, which probably explains why only about 10 percent of people who need them make the purchase. But here's a tip to saving. A built-in charge for long-term service is often bundled into your initial cost. When you purchase hearing aids, ask your specialist if you can unbundle the price and pay only for service if it is required. Since hearing aids are often replaced every four years, you may see significant savings.

Alternatively, good hearing aids can now be purchased online for about half the usual retail cost. Like all hearing aid purchases, online hearing aids may be returned for a refund within one or two months if you don't like them.

Also, keep in mind that hearing aid manufacturers invest a lot of money on research to improve hearing aids, but not everyone requires the latest innovations. Depending on your needs, simplified hearing aids may work as well or better than complicated ones, just as we may find with automobiles, cell phones, and computers.

MYTH *Children younger than twelve months old cannot use hearing aids.*

FACT Not true. Children may use hearing aids by one month of age. When an infant has a hearing loss, it is important to use amplification as soon as possible to minimize delays in language development.

MYTH *Cochlear implants do not work for children who are born deaf or for adults with nerve deafness.*

FACT Not true. The best results with cochlear implants for many children who are born deaf occur when implantation is done at or prior to twelve months of age. Cochlear implants are also very successful in adults who have become deaf later in life.

MYTH *People with cochlear implants hear only buzzing and clicking noises.*

FACT Not true. Cochlear implants are designed to help people understand speech. The average cochlear implant recipient comprehends about 80 percent of what is said and many understand 100 percent in quiet surroundings. Users recognize the voices of friends and relatives and many appreciate music. Nonetheless, hearing achieved with a cochlear implant is not entirely normal, and background noise can interfere with comprehension. Worse results are expected in babies born deaf who do not receive an implant by twelve months of age, people who primarily communicate with American Sign Language, and older adults who also have memory loss or dementia.

MYTH *Most dizziness is caused by stroke or other brain disorder.*

FACT Not true. Most cases of dizziness and vertigo are caused by ear disorders, including the most common cause of vertigo, benign positional vertigo (see chapter 10). Other common inner ear balance disorders include Ménière's disease and labyrinthitis. Disorders of the cardiovascular system are the next most common, followed by brain disorders.

II | Common Ear Problems

4 | Otitis Media

■ Joey's mom was really worried. Joey had been treated with four antibiotics in the past six months. Now his fever was 101 degrees. Last night, green pus drained out of his ear onto his pillowcase. Joey's pediatrician had recommended they see an ear specialist, and now Dr. Johnson, the specialist, was suctioning the pus from the ear. He let Joey's mother look through the ear microscope, and she could see the eardrum had a small hole in it. Dr. Johnson said it would heal itself. When Joey returned a week later, the perforation was indeed gone—but the fluid remained. Dr. Johnson did a hearing test that showed Joey had some hearing loss that would probably be temporary. ■

Otitis media (OM) is an inflammation of the middle ear. The middle ear is a small space that is normally filled with air (see chapter 1) and is lined by a thin mucous membrane. The middle ear is also connected to the airway system that includes the lungs, trachea, nose and sinuses, Eustachian tube, and middle ear. All of the airway structures contain air and are lined with a continuous mucous membrane from top to bottom. Because of the connections between the parts, infection of one part can spread to another. For example, infections like sinusitis can spread downward to the lungs or upward to the middle ear.

Otitis media is the most common childhood illness, causing some twenty million doctors' office visits per year. It has been estimated that 90 percent of children have at least one episode of OM before age two and 30 percent of children have at least six episodes by age seven. Otitis media

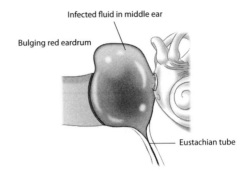

Infected fluid in middle ear

Bulging red eardrum

Eustachian tube

Figure 4.1
In otitis media, the middle ear space is filled with fluid that blocks sound waves.

occurs most frequently in babies between six and eighteen months of age, and there is another peak incidence around the time children enter large day care facilities or preschool. Children younger than age six get otitis media much more often than older children. If the first infection occurs in the first year of life, many more usually follow.

Eustachian Tube Dysfunction

The Eustachian tube (ET) is a narrow connection between the middle ear and the nose and throat. As noted in chapter 1, it is lined with mucous membrane and is normally in the closed position. Specialized muscles open it automatically during swallowing, yawning, or chewing. Or it can be opened consciously by blowing air pressure back into the nose (as we do when we clear our ears for landing in an airplane).

The Eustachian tube is shorter and more nearly horizontal in young children than in adolescents and adults. These two characteristics tend to allow infection or stomach acid to enter the middle ear and cause inflammation of the lining (mucous membrane). The mucous membrane of the Eustachian tube can also become inflamed by inhaled allergens (such as pollen) and by milk that enters the middle ear when a bottle-fed infant drinks while lying on his or her back.

When the Eustachian tube is inflamed, it tends to swell and close more tightly. Mucous from the middle ear may then stop draining and begin to build up, leading to an infection. As a child matures, the Eustachian tube becomes longer and more vertical, and ear infections become less common.

Many other factors play a role in the frequency of otitis media in young children. These include immaturity of the immune system, lack of prior exposure to different strains of viruses and bacteria, inhalation of secondary smoke, and contact with other children who have respiratory infections (runny nose, cough, or sneezing).

Types and Symptoms of Otitis Media

Inflammation of the middle ear takes several forms and can be categorized by the appearance of the eardrum, how long the problem lasts, and complications that may occur. The diagnosis is made by looking through a magnified otoscope or ear microscope while using a pressure bulb. A small puff of air from the pressure bulb will move the eardrum slightly if no infection is present. If fluid is present in the middle ear, the eardrum will not move. The combination of the appearance and movement of the drum are used to make the diagnosis. The most common types of OM are acute otitis media and otitis media with effusion (see below). The other categories can be thought of as progressive stages of complications.

Acute Otitis Media

Acute otitis media (AOM) is usually a bacterial infection that is associated with a rapid onset of pain and temporary hearing loss. The child may also be irritable and have a fever and headache. On examination, the eardrum looks red and often bulging because of the infected fluid buildup behind it. It does not move in response to a slight puff of air. AOM can also be caused by viruses and other less common disorders.

Otitis Media with Effusion

Otitis media with effusion (OME) is a buildup of fluid behind the eardrum without signs of bacterial infection. It may be the result of blockage of the Eustachian tube due to allergy, irritation, or acid reflux. Otitis media with effusion may also be the first or last stage of acute otitis media.

Children with cleft palate or Down syndrome are at higher risk for developing otitis media with effusion because of differences in their anatomy. In addition, children from lower socio-economic families are more likely to have both AOM and OME. This is thought to be correlated with a higher incidence of parental smoking, the use of bottle feeding, and crowded conditions in group day care or the home. Otitis media with effusion is not associated with rapid onset of pain or fever; instead, the symptoms are hearing loss, fullness, and, sometimes, imbalance.

Children who ask for a louder volume on the TV, have difficulty understanding instructions, pull on their ears, or display lethargy or irritability may actually have OME. Sometimes there are no symptoms so it is up to the doctor to order hearing tests and follow-up on anyone who has had acute otitis media to be certain that the fluid resolves after treatment.

Otitis media with effusion may begin with thin watery fluid, but if it persists over time the fluid thickens and turns yellowish. When the fluid reaches a gel-like or even semi-solid consistency, the condition is called *chronic otitis with effusion*, or sometimes "glue ear." At this point, the thick fluid is unlikely to resolve even with medications or further waiting.

OME often causes conductive hearing loss by blocking sound waves, and if the condition is not controlled, may lead to multiple recurring episodes of acute otitis media. If allowed to persist for a period of more than three months, OME can lead to language acquisition delays, reading delays, degeneration of the eardrum, and other structural complications. The first few years of life are a crucial time for children to learn language by hearing other people talk. If a young child has even a mild hearing loss, he or she may miss parts of words and that can interfere with language development.

Diagnosis

Medical Examination

The key to the diagnosis is the appearance of the eardrum, but getting a good look at it is often not easy. Infants may have tiny, collapsible outer ear canals, often filled with vernix (the creamy coating from uterine life). Older children may have wax obstructing a full view. Crying in response to being restrained can cause a normal eardrum to turn red (just as the face turns red when a child is crying). Young children and infants generally allow the doctor just a momentary glance before they move. It may become necessary to refer a child to an ear specialist just to get a good look and diagnosis.

Testing

The most useful test for OME is tympanometry (see chapter 26). Tympanometry is used to measure the movement of the eardrum. It is very sensitive to the presence of fluid, stiffness, and position of the drum and middle ear and, in the authors' experience, it is accurate about 90 percent of the time. This test is done by inserting a probe into the outer ear canal of a child for five seconds or so. If the ear canal is completely blocked, for example by wax, tympanometry will not work. The hearing is also tested to determine the type and degree of loss. Most typical is a mild to moderate conductive loss (see chapter 26).

Treatment

Efforts to prevent OM are safe, effective, and important. The routine vaccination of all children against *Streptococcus pneumoniae* and *Haemophilus influenzae* is very helpful, since most episodes of AOM are caused by these two bacteria. Other preventative factors are living in a smoke-free environment, avoiding large day care facilities early in life, nursing for six months, and using a proper nursing position (not flat on the back).

Acute Otitis Media

Unlike throat cultures, which can be easily performed with a swab, cultures of the ear for AOM are more invasive (requiring a needle puncture of the eardrum). Consequently, they are not routinely performed. However, it has been found that 90 percent of AOM can be traced to three bacteria. *Streptococcus pneumoniae*, *Haemophilus influenzae*, and *Moraxella catarrhalis* are the most common. Together they cause over 90 percent of AOM. Therefore, antibiotic treatment, when indicated, is tailored to be effective against these three.

However, in newborns and infants less than six weeks of age, more resistant bacteria can also cause AOM and must be accounted for. Some newborns get otitis media before they are discharged from the hospital. That means the bacteria may be resistant to usual treatment. For newborns and infants, the infection is often cultured before treatment.

Viruses play an important role by causing inflammation and mucous membrane damage that allow bacteria to gain a foothold. The most common are respiratory syncytial virus, influenza, rhinovirus, and adenovirus.

Discussions are ongoing concerning withholding antibiotics in routine cases of acute otitis media as well as which antibiotics to use if necessary. As of 2013, a committee of the American Academies of Pediatrics and Family Practice made several recommendations (table 4.1).

Nonetheless, some controversy exists, especially about withholding antibiotics and using the largely ineffective antibiotic amoxicillin. Randomized, prospective, placebo-controlled studies show that antibiotics are highly effective in treatment of AOM and cured study subjects two weeks faster than non-antibiotic treatment.

And although amoxicillin is least expensive, it does not provide the best coverage. Of the three bacteria that cause 90 percent of AOM, *H. influenzae* is up to 50 percent resistant and *M. catarrhalis* is 90 percent resistant to amoxicillin.

Use of an antibiotic with such high levels of bacterial resistance can result in unnecessary trips to the doctor's office and avoidable expense. Resistance varies by community, and individual physicians will know best what works for patients in their community. (The authors of this book fa-

---

Table 4.1
Recommendations for Physicians Treating Acute Otitis Media (AOM) from the American Academies of Pediatrics and Family Practice

- Avoid using antibiotics in mild cases
- Prescribe amoxicillin as the antibiotic of choice
- Re-evaluate children within seventy-two hours if not treated
- Encourage breastfeeding for the first six months
- Give *Pneumococcal* and *H. influenzae* vaccinations

Simplified and summarized from the American Academy of Pediatrics Clinic Practice Guideline, *Pediatrics* (2013 Mar); 131(3).

vor individualized treatment of all patients rather than one-size-fits-all medical practice.) Ventilation tubes are also used when multiple episodes of AOM occur within a short period of time in spite of multiple courses of antibiotics.

Otitis Media with Effusion

About two of every three patients with AOM (active infection) develop OME (residual fluid behind the eardrum). The residual fluid will eventually resolve on its own in most children. As a rough rule of thumb, 30 percent resolve in thirty days, 60 percent resolve in sixty days, and 90 percent in ninety days. If fluid is still present at three months, chances are high that the middle ear effusion will not resolve on its own. While conservative practice, in the authors' opinion, requires children to wait for three months to see if the fluid will clear up on its own, most adults with the same disorder demand the fluid be removed by one month owing to discomfort and hearing loss. Unfortunately, medical treatments for OME, including antihistamine-decongestants, steroids, and allergy treatment, have no measurable benefit compared to placebo.

Ventilation tubes are often recommended when fluid is present and causes hearing loss in both ears for three months. If OME begins to cause damage by weakening and collapse of the eardrum itself, ventilation tubes may become necessary sooner in order to prevent long-term damage.

Ear Tubes

Ear tubes (also called tympanostomy tubes, pressure equalizing tubes, ventilation tubes, etc.) are small grommets placed through a tiny incision in the eardrum. The tubes are made of a variety of plastics, Teflon, or stainless steel and work by holding the incision open for six to eighteen months before they are extruded by the body. Tubes allow middle ear fluid to escape rather than build up inside the middle ear. Tubes also permit antibiotic drops to enter the infected middle ear space along with air to help the mucous membranes to return to normal. There are several widely accepted indications for placing tubes (table 4.2).

Complications with tubes include perforation of the eardrum, drainage from the ear, obstruction, and early or delayed extrusion of the tubes.

Table 4.2
Indications for Middle Ear Tube Placement
Acute otitis media (AOM)
Three infections in six months
Four infections in twelve months
Otitis media with effusion (OME)
Fluid present for three months in both ears
Fluid present for six months with degeneration in one ear

With standard tubes, there is a 2 percent chance that the opening in the eardrum may not close on its own. With so-called long-lasting tubes, the residual perforation rate can approach 16 percent. The tube may not extrude on its own and may have to be removed after three or four years of careful observation (2 percent of cases). Some doctors suggest removal by two years, but in our experience, this may lead to the need for reinsertion of tubes a few months later. If drainage through the tube occurs and is not treated with ear drops, the drainage may solidify and form a crust that blocks the tube.

It was once thought that children with tubes should avoid swimming. However, it is safe for most children to swim without ear plugs or headbands in private, adult pools. Avoid submerging the ear in soapy bath water, hot tubs, and unfiltered or non-chlorinated kiddie pools, freshwater lakes in hot climates, at the beach, and in large public pools. Swimming in the ocean, away from the beach, is permitted because the water is less contaminated and does not contain sand and seaweed particles.

Adenoidectomy

If middle ear infections or effusions return after the first set of ear tubes are extruded it may be wise to replace the tubes in combination with removal of the lymphatic tissue called the adenoids. The adenoids normally act as a trap for bacteria, but after too many infections they may become a reservoir and lead to new infections. Adenoidectomy is a somewhat more risky and painful operation than insertion of ear tubes. There is no reason for someone to have their tonsils removed—to have a tonsillectomy—to treat otitis media, because that surgery does not help prevent or cure ear infections.

Complications of Otitis Media

Chronic Suppurative Otitis Media

Chronic Suppurative Otitis Media (CSOM) is a bacterial infection of the middle ear (AOM) that lasts more than six weeks. To make this diagnosis, some ear specialists also require evidence of drainage of infected material through a perforation of the eardrum. Perforations are caused by infected material under pressure in the middle ear. Most of the time, perforations will close after the pus drains out of the middle ear. But if the perforation persists, you will be referred to an ENT for more comprehensive diagnostic evaluation and treatment.

Adhesive Otitis Media

If fluid remains in the middle ear more than a few months, a series of progressive degenerative changes begin to affect the eardrum. The first change is thinning or atrophy of the eardrum. This is a result of a combination of enzymes in the middle ear fluid and negative middle ear pressure.

Eventually the eardrum collapses (atelectasis of the eardrum). In this case, the middle ear space is mostly gone and the eardrum lies directly on the cochlea and bones of hearing. If it remains there, it will eventually attach to those underlying structures (adhesive otitis), at which time the changes are usually irreversible without major surgery and cause moderate to severe hearing loss.

Spread of Infection

The most frequent complications of otitis media, as noted, are hearing loss, speech, language and reading delay, and degenerative changes of the eardrum. But infectious complications, while less common, are more threatening.

Infectious complications are generally categorized by their location and duration. Infections of the middle ear tend to spread to adjacent air spaces. This can result in *mastoiditis* (spread to areas behind the middle ear) and *petrositis* (spread to areas deeper than the middle ear). Infection can also involve the facial nerve and inner ear with very serious results.

Rarely, an ear infection can spread to areas around the brain and may cause meningitis, brain abscess, blood clots in major veins, and hydrocephalus. These intracranial complications can be life threatening.

5 | An Overview of Hearing Loss

Blindness separates people from things; deafness separates people from people. —HELEN KELLER

Hearing loss is the third most common health problem in the United States and is becoming more widespread every year. In 1971, 13.2 million Americans self-reported hearing loss. This number rose to 28.6 million in 2001 and was over 37.5 million by 2012.

The most common cause of hearing loss is aging. It is estimated that more than half of people over seventy-five years of age have significant hearing loss. For people between the ages of twenty and sixty-nine, most hearing loss is caused by noise exposure (see chapters 7 and 22). Other causes of hearing loss in adults include medical disorders such as otosclerosis, Ménière's disease, sudden deafness, benign tumors of the auditory nerve, and as a side effect to the use of ototoxic medications such as intravenous antibiotics and anti-cancer treatments.

Hearing loss causes safety issues. When we can't hear warning sounds like the noise of an approaching truck, a fire alarm, or a siren it becomes dangerous to cross a street, much less drive a car. Hearing loss is also responsible for communication disorders in adults, and secondary effects include reduced quality of life, loss of intellectual stimulation, social isolation, depression, economic loss, and acceleration of dementia in the elderly.

The most common cause of hearing loss in children is middle ear infection (otitis media) (see chapter 4). Less common but more severe losses in children are caused by hereditary (genetic) abnormalities (see chapter 6). Untreated hearing loss in children results in speech and language disabilities, cognitive and reading impairment, social and emotional maladaptation, and learning disabilities.

In our lifetimes, most of us will suffer some degree of hearing loss. Early diagnosis and management can reduce or even resolve the communication, social, intellectual, and safety issues associated with hearing disability. Methods of treatment include hearing aids, medical treatment, surgery, language therapy, and cochlear implants, each described in subsequent chapters.

Types of Hearing Loss

Hearing loss is a partial or complete inability to sense and understand sound that can result in disability. It is described in specific ways including type of hearing loss, age of onset, genetic status, and severity.

Hearing loss is often divided into two general types: *conductive* (caused by obstruction of sound waves) and *sensorineural* (a result of damage to the nerve or other structure of the inner ear). Mixed hearing loss refers to a combination of both. Sensorineural hearing loss comes from the inner ear and conductive hearing loss comes from the middle or outer ears (see figure 5.1).

Nerve Deafness (Sensorineural Hearing Loss)

Sensorineural hearing loss (SNHL) is most often the result of damage to hair cells in the cochlea (see chapter 1). Hair cells receive sound vibrations and convert them into nerve signals that travel to the hearing centers of the brain. This type of hearing loss is permanent since there is currently no repair available for human hair cells and they cannot be replaced. Damage to hair cells may be caused by the aging process (see chapter 7), exposure to loud sound (see chapter 22), infection, inflammation, and head injury. There are also certain medical conditions (hereditary deafness, meningitis, Ménière's disease, etc.) and even prescribed medications (see chapter 24) that cause sensorineural hearing loss.

Sensorineural hearing loss also can be caused by damage to the nerve of hearing, for example by benign tumors (acoustic neuroma and meningioma) or nerve infections. Examples of nerve infection include cytomegalovirus, rubella, meningitis, and syphilis. Traumatic head injury may also cause sensorineural hearing loss by tearing the cochlear nerve or causing inner ear concussion. Nonetheless, sensorineural hearing loss caused by damaging the nerve of hearing is not common.

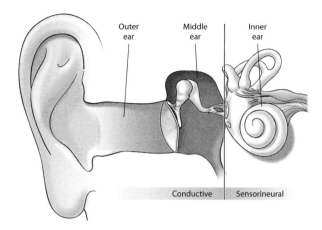

Figure 5.1
Conductive hearing loss is caused by disorders of the outer and middle ear. Sensorineural hearing loss is caused by disorders of the inner ear.

Hearing loss coming from the brain is even less frequent. It may be caused by developmental malformation, jaundice (yellowish discoloration), and viral infections in newborns. Bacterial meningitis can occur at any age, but is more common in childhood. It is a life-threatening infection of the membranes (called meninges) and fluids that surround the brain. Even with early diagnosis deafness can result. Sometimes doctors will use steroids along with antibiotics to reduce hearing loss associated with meningitis.

Conductive Hearing Loss

The other major category is *conductive hearing loss.* In this type, sound waves are blocked from passing through the outer or middle ear by some mechanical obstruction. If the sound waves are blocked, they can't stimulate hearing.

People with conductive hearing loss are not completely deaf and understand speech when it is loud enough. Conductive hearing loss is usually accompanied by a feeling of fullness in the ear and by an echo or extra loudness of one's own voice. This is called the occlusion effect, a phenomenon that anyone can sense by blocking one ear canal and speaking normally. The voice will be emphasized in the blocked ear. Since conductive hearing losses are mechanical, hearing can often be permanently restored by removing foreign material from the ear canal, repairing the eardrum, or replacing abnormal bones of the middle ear. Hearing aids are also effective in treating conductive hearing loss.

Wax buildup and outer ear infections are the most common causes of outer ear conductive hearing loss. Obstruction can also be caused by foreign bodies in the ear canal, malformation of the ear canal, benign growths, or, rarely, even cancer (see part 3). These obstructions act like ear plugs to keep the sound waves from getting inside.

The most common causes of middle ear conductive hearing loss are infections (acute otitis media) and fluid in the middle ear (otitis media with effusion) from allergies, colds, or blocked Eustachian tubes. Other causes are perforations of the eardrum, fusion or disruption of the bones of the middle ear, and benign growths of the middle ear.

Another reason for conductive hearing loss is otosclerosis, a common disorder caused by bony overgrowth of the stirrup (stapes), the third bone of hearing in the middle ear. New bone formation and overgrowth fuses this bone to the surrounding inner ear bone, preventing normal vibration. Otosclerosis can run in families and often begins around the age of thirty. It can be treated with hearing aids or surgery (stapedotomy, see chapter 19).

Damaged or malformed middle ear bones can also cause conductive hearing loss. The three tiny bones in the middle ear can be damaged by infection or head injury. In rare cases, the bones do not develop properly in the second trimester of gestation. The ossicles can usually be repaired or replaced surgically to restore hearing.

Holes in the eardrum cause conductive hearing loss and can result from severe otitis media, head injury, blast injury, or punctures from hard objects like cotton swabs or bobby pins. The perforations normally heal by themselves and the hearing loss is temporary. If spontaneous healing does not occur, the eardrum can be repaired surgically, often with restoration of hearing (see chapter 18).

Severity of Hearing Loss

Threshold

A hearing test (audiogram) can determine the severity of hearing loss. The amount of hearing loss is measured in decibels (dB) and classified into the following categories: mild, moderate, severe, and profound. Normal hearing levels are 0–25 dB in adults and 0–19 dB in children. Children need better hearing to learn language than adults need to use the language they have already learned.

The categories of hearing loss are defined by the softest sound (threshold) that someone can hear:

- *Mild hearing loss* (26–40 dB) causes difficulty hearing soft sounds and understanding conversation in noisy backgrounds. Children with mild hearing loss may require hearing aids and/or preferential seating in class in order to develop language normally.
- *Moderate hearing loss* (41–70 dB) causes difficulty with moderate and loud speech. Hearing aids are needed by children and adults.
- *Severe hearing loss* (71–90 dB) impedes learning and communication even with hearing aids. Some people with severe loss (or even "moderate to severe hearing losses") will need a cochlear implant in order to understand the words they hear.
- *Profound hearing loss* (greater than 90 dB) generally requires a cochlear implant or signed language communication.

Most of the time, the high frequency loss is greater than the low frequency loss which forms a sloping pattern on the hearing test. For example, a person may have a mild hearing loss in the lows and a moderate loss in the highs. This is called a mild to moderate hearing loss.

Speech Understanding

The ability to understand speech is the second, equally important, way to describe hearing loss. We measure speech recognition or speech perception as the percentage of words that can be understood. In speech recognition tests, recorded words or sentences are delivered at a loudness that is adequate to hear the sound. Speech understanding scores of 100 percent are perfect and less than 60 percent can be disabling. Older adults and those with certain ear disorders may hear sounds adequately but not be able to understand the words. In such cases, imaging tests like an MRI (magnetic resonance imaging) may be necessary to rule out the presence of a tumor or other neurological problem.

Two Common Causes of Nerve Deafness in Adults

Age-Related Hearing Loss

Loss of hearing due to the aging process is a growing problem—people are living longer and yesterday's baby boomers have become today's elders. It is the most common neurodegenerative disorder and cause of communication failure in the United States.

The severity and age of onset of age-related hearing loss (called *presbycusis*) tends to be hereditary and men are disproportionately affected. Age-related hearing loss has a great impact on the quality of life of tens of millions of people and is associated with all forms of dementia (see chapter 7).

Noise-Induced Hearing Loss

Exposure to loud noise has become a common cause of hearing loss, especially in adolescents and adults. For most this is a gradual process, building up over several years without warning signs such as pain. Hearing protection is necessary during high-sound-level activities and the volume of personal music players should be controlled. (For more detail, see chapter 22, Noise-Induced Hearing Loss.)

Hearing Loss in Children

For children, hearing is the key to language development and is critical to learning, socialization, and cognitive development. Children learn communication through imitation. If they can't hear words clearly, they won't be able to say them clearly. Later on, they will have difficulty reading and spelling. When even mild hearing loss is unrecognized and untreated in infants and young children, social dysfunction and academic problems with reading and writing occur later in school.

Congenital Hearing Loss

Congenital hearing loss means that the loss was present at the time of birth, or that the condition that would later cause hearing loss was present at birth. For example, some forms of genetic deafness (the gene is abnormal at birth) only begin to show progressive loss of hearing years later. Over 50 percent of childhood deafness is caused by genetic disorders. Congenital hearing loss can occur with other related physical abnormalities (in which case it is called *syndromic*) such as kidney, eye, or heart disorders. Or it can occur with no associated abnormalities (*non-syndromic*). Viral infections, high fever, meningitis, prematurity, jaundice (yellow skin and eyes), and the use of certain life-saving antibiotics are also associated with newborn hearing loss.

Acquired Hearing Loss

Acquired hearing loss means that neither the loss, nor its cause, were present at birth. Examples of causes of acquired loss include otitis media with effusion, blockage of the ear canal with vernix (the creamy substance that covers newborns) or wax, perforation of the eardrum, head injury, excessive noise exposure, and exposures to certain intravenous antibiotics or anti-cancer chemotherapy (see chapter 24). Hearing loss can also be acquired as a result of infections such as measles, mumps, rubella, or meningitis that develop after birth (see chapter 6).

Self-Evaluation for Hearing Loss

If you suspect you have a hearing loss, you probably do. It is wise to get a professional evaluation. But if you want to get a little back-up first, complete the questionnaire in table 5.1 to see where you score.

It is also possible to test your hearing online, although these tests are not always accurate, tend to over-identify hearing problems, and, in some cases, are used as marketing tools to refer potential customers for hearing aids. Just search for "Hearing Test" and you will find dozens.

Table 5.1
Self-Administered Adult Hearing Questionnaire

	No (0 points)	A Little (1 point)	A Lot (2 points)
1 I often ask people to repeat.	☐	☐	☐
2 I find it hard to hear in noisy situations.	☐	☐	☐
3 I hear words but I do not always understand them.	☐	☐	☐
4 I avoid groups because it is hard to hear.	☐	☐	☐
5 I turn up the television too loud for others.	☐	☐	☐
6 I get frustrated that my spouse mumbles.	☐	☐	☐

Scoring   0–2   Probably okay at this point.
                 3–6   Time to get a formal hearing test.
                 7–12 Watch out for that train!

6 | Hearing Loss in Children

■ The first thing Dad did when Meredith was born was to check out the little details. Fingers, toes, eyes, ears, lips—everything perfect. So he and Mom were both shocked when Merrie did not pass her newborn hearing screening the next day. The nurse told them that it may mean nothing at all, but she advised them that Merrie needed further testing. After a series of tests Merrie's ear specialist told Mom and Dad that their perfect child had a nerve-type hearing loss. It would most likely be permanent and could possibly get worse. This was Dad's worst fear because hearing loss ran in his side of the family. ■

Hearing loss is the most common birth defect in the United States, affecting one to three of every thousand newborns. Some 25 percent of children who are born with impaired hearing will have severe to profound loss, enough to prevent normal language acquisition and intellectual, emotional, and social development. Figure 6.1 shows the relative incidence of the most common birth defects. Notice that hearing loss is by far the most common.

The impact of untreated congenital deafness is high, averaging well over $1 million per profoundly hearing impaired child. This figure includes direct costs such as education (a third of total cost) and loss of earning potential (two-thirds of total cost).

Causes of Congenital Hearing Loss

Congenital means that the hearing loss is either present at birth or the genes that will cause deafness later are present at birth. For example, in

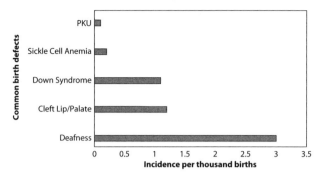

Figure 6.1

Incidence of birth defects per thousand births in the United States. Hearing loss is the most common birth defect in the United States. It occurs more than twice as frequently as any other birth defect.

genetic deafness from Alport syndrome (kidney failure and hearing loss), the abnormal Alport gene is present at birth but hearing loss may not be noticeable until adolescence. Non-genetic congenital deafness can be caused by viral infections, bacterial meningitis, newborn jaundice, low birth weight, low oxygen levels, some life-saving antibiotics, and other causes (see figure 6.2).

Genetic Hearing Loss

Inherited deafness is complex and the following explanations are just the basics. Genetic abnormalities are responsible for about 60 percent of

Figure 6.2

Causes of congenital sensorineural hearing loss. The majority of worldwide hearing loss in newborns is caused by genetic inheritance.

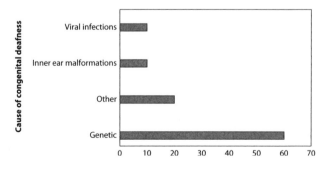

congenital hearing loss in children. Over a hundred different gene mutations causing deafness have been identified. Everyone has about twenty-four thousand genes that are grouped together on chromosomes. Half of our forty-six chromosomes come from each parent. As described below some forms of deafness only occur when the infant inherits abnormal hearing genes from both parents (*recessive* deafness) and other forms require only an abnormal hearing gene from one parent (*dominant*).

Syndromic vs. Non-Syndromic Deafness

The term *syndromic* indicates that deafness is associated with other clinical abnormalities. Examples include renal failure with deafness (Alport syndrome); widely spaced, different colored eyes and a streak of gray hair with deafness (Waardenberg syndrome); goiter with deafness (Pendred syndrome); and heart problems with deafness (Jervell and Lang-Nielson syndrome). In all, about 20 percent of congenital hearing loss is syndromic and over four hundred syndromes have been identified.

Conversely, children with non-syndromic deafness have no other abnormalities. Non-syndromic deafness is much more common, accounting for about 80 percent of congenital hearing loss.

Dominant vs. Recessive Hearing Loss

Genes are categorized as dominant (20 percent) or recessive (80 percent). Dominant genes can be thought of as being so powerful that only one (from one parent) is required to cause hearing loss. Most syndromic hearing loss is associated with dominant genes. On the other hand, two recessive genes (one from each parent) are required to cause recessive hearing loss.

For a child to be born deaf, he or she must inherit at least one dominant deafness gene (from one parent) or two recessive deafness genes (one from each parent). If one parent has a dominant deafness gene, that parent will be deaf and has a 50 percent chance of producing a deaf child. If both parents are dominant, they have a 75 percent chance of producing a deaf child.

In the most common situation, both parents are hearing but each carries one recessive deafness gene. Parents who have a single recessive gene have normal hearing and are called *carriers*.

It may be difficult to trace deafness from one generation to the next in families with recessive genes. If both parents are carriers (one recessive gene each), the chance of having a child who is hearing impaired is 25 percent; the chance of having a child who is a carrier is 50 percent; and the chance of having a child with no recessive deafness gene is 25 percent. Over 90 percent of deaf children are born to two normally hearing parents and over 95 percent are born to at least one hearing parent.

Rare Forms of Inheritance: X-linked and Mitochondrial

These two types of inheritance are transmitted solely by the mother. X-linked deafness affects predominantly males and mitochondrial deafness affects both males and females. But a male with mitochondrial deafness cannot pass it on. Together these two rare forms account for only about 2 percent of genetic deafness.

Genetic Counseling and Testing

Genetic testing is just one part of the process of genetic counseling, in which potential parents or parents concerned with inherited deafness are advised about the cause of their hearing loss or the chances of developing or transmitting it. The information is complex and may have a large impact on family life, so the counseling aspect is as important as the actual test results.

Although blood tests can identify over a hundred genes that have been associated with hearing loss, many of those tests are limited to research laboratories. However, the most prevalent gene abnormalities are *GJB2* and *GJB6*. Clinically this is often referred to as connexin-related deafness and readily available testing for these two genes identifies the cause of 50 percent of childhood deafness.

Acquired (Non-Genetic) Congenital Deafness

Infections

Most congenital infections are viral and cause an estimated 10 percent of congenital deafness. Infection can be transmitted by the mother before birth through the placenta, during birth from infected vaginal secretions, after birth from breast milk, or from airborne germs transmitted by coughing and sneezing. The severity of the infection and the amount of hearing loss caused are related to the type of virus as well as the stage of maturity of the child at the time of infection. A viral infection in the first trimester may cause more damage than the same virus would cause in the third trimester.

The most frequent viral infections include cytomegalovirus (CMV), rubella (German measles), and herpes simplex. Mothers can also transmit toxoplasmosis (a parasite) and syphilis (a spiral bacterium) that cause hearing loss. You may come across the acronym TORCH. This is a group of blood tests that screen for infections in newborns. The screening detects toxoplasmosis, rubella, cytomegalovirus, herpes simplex, HIV, and syphilis infections.

Congenital syphilis may cause deafness, saddle-nose, and deformed teeth, fingers, and toes. It may also result in miscarriage, prematurity, and stillbirth. Adult-onset deafness and eye inflammation may also occur from a delayed form of congenital syphilis.

Malformations of the Inner Ear

Up to 20 percent of children with congenital hearing loss have a cochlear malformation. Abnormal genes causing some malformations have been identified. Inner ear malformations are associated with a risk of meningitis.

The most common inner ear malformation is called *enlarged vestibular aqueduct* (EVA), which can cause a progressive loss of hearing during childhood and adolescence. The progressive loss usually occurs in steps that may follow mild head bumps. Some ear specialists recommend that children with EVA avoid contact sports and rough play. The vestibular aqueduct is a bony channel that contains inner ear fluid and it is enlarged in up to 5 percent of children.

Mondini malformation is an enlarged vestibular aqueduct plus incomplete inner walls (partitions) of the cochlea. Both enlarged vestibular aqueduct and Mondini deformities may be associated with a genetic abnormality that also causes Pendred syndrome (deafness and goiter).

Ototoxic Drugs

Sometimes life-saving antibiotics are required for pregnant mothers or infants even though side effects include hearing loss. Infectious disease specialists, obstetricians, and pediatricians use these drugs only as a last resort. They monitor blood levels to minimize chances for damage, and they communicate the possible risks to families. The most common medications to cause hearing loss are a family of antibiotics known as aminoglycosides. This family includes gentamicin, tobramycin, kanamycin, and streptomycin. There is a genetically determined high risk in some families. Tests are available in such cases.

Diuretics ("water pills" used to treat high blood pressure or fluid retention) can also be ototoxic. The loop diuretic ethacrynic acid is ototoxic and is used only when a similar drug, furosemide, cannot be prescribed because of a sulfur allergy. Ototoxicity is more likely when aminoglycosides are used in combination with loop diuretics.

Hearing loss can also be produced by cancer chemotherapy. The most likely drugs to cause hearing loss are cisplatin, carboplatin, and oxaplatin. Good hydration, slow administration, and use of lower dosage can all play a role in reducing toxicity. Vincristine can cause reversible hearing loss and is less toxic but often less effective. Both oral and in-the-ear steroids have been used to reduce ototoxicity. Other ototoxic medications include high-dose aspirin and other NSAIDs (nonsteroidal anti-inflammatory drugs) and quinine (all reversible when medication is discontinued). In addition, mercury and lead poisoning are associated with hearing loss. (See chapter 24 for more detail on ototoxic medications.)

Newborn Jaundice

Hyperbilirubinemia, or too much bilirubin, is a common condition of infants that often requires medical treatment. It is caused by an excessive buildup of bilirubin, a blood chemical that occurs when the body replaces old blood cells with new ones. Jaundice (yellowness of the skin, eyes, and other tissues) is a result of the pigment of the bilirubin.

When abnormal levels reach about 13 mg/dL, treatment is required with phototherapy (exposure to blue fluorescent light reduces unconjugated bilirubin). Jaundice that requires phototherapy is severe enough to cause hearing loss. The wavelength of this light has the property of breaking down bilirubin so that it can be excreted in urine or stool. In severe cases of hyperbilirubinemia, exchange transfusions (replacing much of the newborn's blood with donor blood) are necessary.

Temporary Causes of Congenital Conductive Hearing Loss

Some newborn hearing loss is caused by temporary conditions that resolve spontaneously or with treatment. Newborns have tiny outer ear canals that tend to collapse during hearing testing. Collapse can lead to false positive test results. The canals can also be filled with vernix, the white, creamy covering of fetal skin that has antibiotic and moisturizing properties. Both conditions can cause a temporary hearing loss that tends to resolve in the first weeks of life. Amniotic fluid may also be swallowed by the fetus and end up in the middle ear. Acute otitis media is an infection of the middle ear that may rarely occur soon after birth. It causes fluid to build up in the middle ear and blocks sound waves from entering.

Identification of Hearing Loss in Infants

Early identification of newborn hearing loss is essential to allow treatment that can reduce the linguistic, intellectual, social, and emotional deficits caused by deafness. Two common methods of early identification include universal newborn hearing screening and identification of high-risk factors for hearing loss. Best results occur when both methods are applied.

Universal Newborn Hearing Screening

Newborn hearing screening is a public health program designed to identify infants with hearing loss and enhance timely referral to hearing health care professionals. According to the CDC, in 2011, 98 percent of American newborns underwent hearing screening in the first few months of life.

Ideally newborns should be screened before they leave the hospital but no later than one month of age. There are two methods of screening: the *auditory brainstem response test* (ABR) and the *otoacoustic emission test*

(OAE). Both tests are very sensitive and are structured to identify problems. But they are not highly specific, falsely identifying hearing loss when there is none in up to 30 percent of infants. As a result children with normal hearing are too often referred for unnecessary and expensive batteries of diagnostic testing.

The ABR test measures the electrical hearing signals that run from the cochlea to the mid-brain. The signals are picked up by electrodes taped to the head. The stimulus is a click that is presented through headphones at a loud level and soft level. Testing takes about five minutes if the child is sleeping (longer if the child is not lying still) and may be performed by a health professional or trained lay volunteer.

The OAE test measures sound waves that are created in the cochlea during the process of hearing. These emissions travel back out to the ear canal where they are picked up by a tiny microphone. Unfortunately, this test does not identify problems with the nerve of hearing or the hearing areas of the brainstem. It is also more likely to be affected by temporary fluid in the middle ear and ear canal causing more unnecessary trips for testing after discharge from the hospital.

High-Risk Factors for Newborn Hearing Loss

A *high-risk register* (HRR) identifies infants who are at risk to have hearing loss. These risk factors are presented in table 6.1. High-risk registers are not as sensitive as universal newborn screening and will miss about 50 percent of infants who have hearing loss. On the other hand,

---

Table 6.1
High-Risk Factors for Newborn Hearing Loss

- Admission to newborn ICU for more than five days
- Prematurity (gestation of less than thirty-seven weeks)
- Maternal infection
- Abnormal appearance (head, face, ears, neck, eyes, fingers)
- Ototoxic drugs given to mother or child
- Yellow skin or eyes (hyperbilirubinemia)
- Identification of known deafness syndrome

---

certain types of congenital hearing loss have a delayed onset and cannot be identified with hospital-based screening tests. If either the screening test or HRR show abnormality, the infant is referred for definitive evaluation by a team that consists of an audiologist and physician.

When high-risk factors are identified, the child should be referred for complete evaluation of hearing even if he or she passes newborn hearing screening tests. Ideally the complete evaluation would be performed at a facility that offers a team approach to hearing loss including an otolaryngologist, audiologists, and a language specialist.

Since hearing loss may not be present until later in life, ongoing evaluation throughout childhood is important. One method is surveillance of milestones by physicians and patients of communication development between two months and two years of age. Children who do not meet these milestones should be referred for hearing and language evaluations.

Table 6.2
Normal Hearing and Language Milestones

0–3 months
- Startles to loud sounds
- Quiets when spoken to
- Changes sucking behavior in response to sound

3–6 months
- Identifies direction of sound with eyes
- Listens to music

6–12 months
- Turns head toward sound
- Recognizes first words
- Begins canonical babbling (rhythmic, repetitive)

12 months
- Says first words

24 months
- Follows one-stage commands ("Find Daddy.")
- Speaks in short phrases ("Mommy come.")
- Knows dozens of words

Current strategies to identify hearing loss that does not begin until children are older include awareness training for parents and teachers and periodic testing. Hearing is routinely tested through public schools or physician offices at ages four, six, ten, and eighteen. Classroom teachers are also trained to recognize hearing loss and to discuss this with parents and the school audiologist.

Evaluation of Hearing Loss in Infants

If your newborn is found to have a hearing loss, the next step is an evaluation by an ear specialist. Your physician will conduct a thorough history and physical examination. Questions focus on birth history (such as: During pregnancy, did the mother have a fever or rash, take any medications, have high blood pressure?) and family history (for example: Does hearing loss run in the family?). During the physical examination the specialist is looking for any abnormalities: Are the ears, head, face, fingers, and toes well formed? Is there anything obstructing the ear canal? Is there fluid or infection of the middle ear? Are the eyes, nose, and throat normal? Testing follows and is guided by the findings of the history and physical examination.

Testing

This evaluation includes hearing tests and may include blood tests, urinalysis, and imaging (CT scan or MRI) of the ear and brain. Certain eye, kidney, and heart abnormalities may occur in conjunction with deafness. For this reason an eye examination by a pediatric ophthalmologist, urine testing, and EKG may all be indicated.

While hearing screening tests are designed to be fast and inexpensive, diagnostic hearing testing is time consuming and expensive because it is more detailed and accurate. Diagnostic testing can show the type and degree of hearing loss and help lead to a specific diagnosis and treatment plan.

Hearing testing is best performed by an audiologist working as part of a team. Typical hearing tests in the first six months of life include immittance, auditory brainstem responses (ABR), and otoacoustic emissions (OAE) (see chapter 26).

Interacting with the Doctor

Your ear specialist team must be sensitive to the family when providing the critical answers to question like: Is my baby deaf? Whose fault is it? How severe is it? What caused it? Will it be permanent? What can be done? It may take several visits to fully understand the diagnosis and to work with your team to form a treatment plan.

The first things you will learn are the extent of hearing loss and what that means to your child and family. Further evaluation will be necessary to pin down the cause. If your child has significant hearing loss, a hearing aid evaluation (see chapter 8) will be recommended. Remember, time is of the essence. Do not delay. Early intervention may make a huge difference in your child's future. Your team can provide this service.

If your child has severe to profound loss of hearing, you will have the options of using sign language, using a hearing aid, or, if a hearing aid is not sufficient, using a cochlear implant (see chapter 9). Most parents like to have the chance to meet and talk to other parents who have been in similar positions and had to make difficult decisions. Your team can help arrange this.

7 | Age-Related Hearing Loss

■ After Grandma died, Papa decided to live in a care facility. But over the past two years, he has gone downhill fast—he has few friends, it's hard for him to hear the TV, he finds it frustrating to talk to family when they visit and even more so when they call on the phone. The director says this happens frequently and suggested checking his hearing. She is going to match him up with some outgoing types and thinks he should see the doctor about hearing aids or antidepressants. ■

Sensorineural hearing loss that occurs in older adults is called *presbycusis*. Loss of hearing due to the aging process is a growing issue because people are living longer and yesterday's baby boomers have become today's seniors. Presbycusis is already the most common neurodegenerative disorder and cause of communication difficulty in America.

Approximately 18 percent of middle-age adults (those forty-five to sixty-four years old) suffer from hearing loss compared to 30 percent of those sixty-five to seventy-four, and nearly 50 percent of those older than seventy-four. The degree and age of onset of presbycusis tend to be hereditary, and men are disproportionately affected. Age-related hearing loss has a great impact on the quality of life of tens of millions of older people and is closely associated with all forms of dementia.

Presbycusis is progressive and affects both ears. The degeneration of hair cells in the inner ear is the primary cause. Other parts of the cochlea along with the auditory nerve and brain are also affected by presbycusis.

It is usually not possible to distinguish the natural effects of aging from other factors such as a lifetime of noise exposure, incidents of head trauma, use of medications, and even a history of ear infections. Presbycusis begins by affecting hearing in the higher frequencies and remains more severe in those frequencies. This is significant because high-frequency tones are important in understanding spoken language. Although the cellular degeneration of presbycusis may begin as early as adolescence, it is slow to progress and usually is not a problem until the sixth decade. The age when presbycusis becomes noticeable often runs consistently within families, indicating a genetic component. Among the ironies of aging, men's hearing tends to become worse at a time when women's voices tend to become softer.

Symptoms of Presbycusis

Slowly progressive loss of hearing is unnoticeable at first. As it becomes worse, we tend to ignore it. At some point, usually in our sixties, presbycusis often becomes more annoying, a source of aggravation. Losing the ability to hear other people is first blamed on them, and we wonder why people mumble, speak so softly, talk to us from other rooms, or turn away when speaking. We try to accommodate to this change, along with many others of course, as part of life. But it's not long before our hearing losses begin to irritate others more than ourselves. Old friends ask, why don't you pay attention to what I say? New friends think we are dull or ignoring them.

People with presbycusis may also become withdrawn, and as loved ones and good friends pass away, it is hard to make new friends. This is especially true in the social activities at extended care facilities. Meals in large noisy dining halls and the bustle of group activities make hearing difficult for anyone. It becomes a serious effort to hear someone's name much less remember it and have a conversation. With presbycusis, life tends to become smaller and smaller and personalities can involute unless something is done.

Another problem of presbycusis is loudness discomfort when listening. Presbycusis is associated with narrow dynamic range, the range of loudness between what is barely heard and what is too loud. A senior with presbycusis may ask a friend to speak up, then surprise her by saying, "You don't have to yell at me!"

As mentioned, presbycusis is partly due to degeneration of the hearing centers in the brain. And hearing loss is frequently found in people with early stages of dementia, also called mild cognitive impairment. It is thought that speech signals become difficult for the brain to process. Speech testing may show a disproportionate loss of understanding compared to threshold levels. In these instances, tests of central auditory processing (see chapter 26), the brain's ability to deal with nerve signals, is evaluated.

It can also be valuable to test central processing when older adults who seem to be ideal candidates for hearing aids instead find the aids make sound louder but not clearer. The reason may not be the hearing aid or even the ear, but rather the brain's ability to process input.

Causes of Presbycusis

Aging changes the cochlea microscopically in four known ways: loss of hair cells and their supporting cells; loss of auditory nerve cells (that carry signals to the brain); degeneration of the stria vascularis (part of the cochlea that maintains chemical and bioelectric balance); and thickening of the basilar membrane (which normally vibrates freely).

One of the rare causes of presbycusis is mutation of a mitochondrial gene (see chapter 6). The abnormal gene is passed from mothers to each of her children. It causes hearing loss by affecting the oxygen supply of the inner ear, increasing the rate of programmed cell death (apoptosis) as well as anatomical changes. Two specific inherited DNA deletions have been identified.

Coping with Presbycusis

At the time of writing, there is no cure for presbycusis or any type of nerve deafness. Chapter 30 introduces regenerative medicine in otology, but that approach is currently in the domain of basic and clinical research—not clinical practice.

Beware of claims made by some makers of ear pills, especially if they are called dietary supplements and thereby are not regulated and do not require their claims to be proven. If a pill could really reverse or prevent hearing loss, the inventor would likely be a Nobel Prize winner.

The first step in dealing with presbycusis is to recognize the problem and see an ear specialist. One of the most common disorders mistaken for presbycusis is a wax impaction that is easily removed at the time of your visit to the doctor. Other treatable causes of hearing loss in older adults should also be ruled out before the diagnosis of presbycusis is made. These include disorders of the outer, middle, and inner ears as well as certain brain tumors. Hearing aids are the primary treatment for presbycusis and cochlear implants are effective when indicated.

There are also many accommodations that can be made to enhance communication with mild to moderate presbycusis. Imagine a conversation of an older married couple:

"Can't you stop mumbling all the time?"
"You never did pay attention to what I say."
"Then don't talk to me when I'm in the other room."
"Then put down the newspaper when I have something to say."
"And get your head out of the refrigerator when you say it."
"And turn down the TV so you can hear me."
"And you need to find a nice quiet restaurant if you want dinner."

Most people with presbycusis can communicate well when they control the surrounding noise. For an important conversation, sit facing each other in a well-lit quiet room no more than six feet apart. Speak up but do not exaggerate your words, rush them, or make them too loud. Reading each other's expressions, hand gestures, and natural lip reading are all helpful. Turn off any machines that make noise, including the TV, stereo, and kitchen appliances. Put away books, newspapers, and other distracting items. In a restaurant, sit with your back to the room so your head blocks some of the noise from reaching your ears. Ask your dinner partner to sit in the corner. This will focus the sound into the corner where your dinner partner is sitting causing him or her to speak louder. Your attention is concentrated into the corner, but theirs is distracted by the people walking by.

What to Expect

As we age, hearing loss continues slowly to get worse, about 1 dB per year. As hearing loss progresses, we need to re-program our hearing aids and eventually get new ones. It is also important to avoid noise exposure, have good control of diabetes if present, and avoid ototoxic medications. The most effective treatment for mild cognitive impairment and memory loss is a healthy diet and exercise. This is more important than ear pills or mind training games.

8 | All about Hearing Aids

For the past several years, everyone I know has been pushing me to get hearing aids. So I went to try some out. Six thousand dollars! Sorry, I don't have that kind of money. Well, the audiologist said, try this pair for $5,000. When we got down to $4,000 my wife wrote a check and I started a one-month trial. After a few adjustments, and a lot of getting used to, I could hear the benefit. But I brought the aids back before the trial period expired and got our money back. After all, she said I'd need to buy a new pair in four years anyway.

The National Institutes of Health (NIH) reports that over thirty million American adults report hearing loss but that only one person in five (20 percent) who could benefit from a hearing aid actually uses one. Hearing aids are battery-powered sound processors that are worn in (or on) the ears. They make sounds louder and often clearer—but not always. The most important parts of a hearing aid are the microphone, the amplifier, the digital processor (computer chip), and the speaker.

How They Work

Microphones pick up sound waves and change them to electrical signals. The amplifier makes the signals stronger. The microprocessor is a miniature computer with many functions: it can reduce background noise, amplify selected tones, and compress sound to fit within your hearing requirements without being too loud. The speaker, a miniature of the speakers on your stereo sound system, projects enhanced sound into the ear canal.

Hearing aids are often classified by their size or where they are positioned at the ear.

- *On the ear* (OTE—also called open fit): hooked over the pinna; are very small and surprisingly cosmetic and have become the most popular style for adults. OTE hearing aids may be open fit (no airtight seal needed) or receiver-in-canal (speaker used instead of tubing).
- *Behind the ear* (BTE): hooked over the pinna; similar to the OTE but quite a bit larger. Most commonly used for children.
- *In the ear* (ITE): placed in the pinna and very noticeable unless covered by hair.
- *In the canal* (ITC): are slightly larger and somewhat visible. For use by adults.
- *Completely in the canal* (CIC): placed deeply into the external auditory canal where they may be invisible from the outside. Difficult for children and the elderly to use.

There is often a trade-off of function versus cosmetic appeal. Smaller aids have a smaller battery, processor, and speaker. In some cases, miniaturization may increase price but reduce performance.

Take the time to examine the batteries before you settle on a brand. Hearing aid batteries need to be changed every four to six days. Make sure

Figure 8.1
Five types of hearing aids, showing their relative size
and position in the ear.

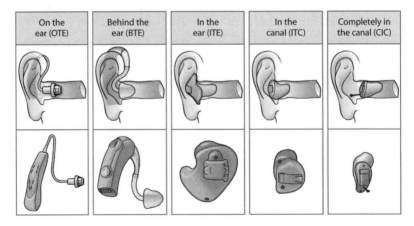

On the ear (OTE)	Behind the ear (BTE)	In the ear (ITE)	In the canal (ITC)	Completely in the canal (CIC)

you can replace the batteries yourself (some are very small and difficult to manipulate) before buying a hearing aid. Ask about the cost of replacement batteries for any device you consider. Some dealers include the price of batteries for one year or more in the purchase price.

Who Should Try a Hearing Aid?

People who have trouble hearing should have an audiogram and hearing aid evaluation. Hearing aid specialists are required to refer children to a physician prior to fitting a hearing aid (and they are strongly advised to refer adults with hearing loss as well). This is to ensure that an underlying disease, which may be treatable if identified properly, is not overlooked. Hearing loss can be caused by medical conditions such as infection, allergy, tumors, perforations, fused middle ear bones, and many other disorders that can be effectively treated.

Keep in mind that hearing aids are imperfect. They assist people with hearing loss but do not cure hearing loss. The fitting process may require adjustments over the first weeks in order to optimize your hearing. It is important to stick with it and accustom yourself to the aid before deciding whether it works or not. In this sense, getting used to a hearing aid is a bit like getting comfortable with contact lenses or bifocals—it takes a little time. If you are considering a hearing aid at the urging of someone else—a spouse, other relative, or friend—chances are you may become easily frustrated by the process.

You may wonder if you should start by buying a hearing aid for only one ear. In the large majority of cases, two hearing aids are better than one. That is because most hearing loss occurs in both ears and because two ears are better than one. Hearing from both ears increases the clarity of sound and reduces background interference, especially in noisy environments.

Buying a Hearing Aid

Try to identify a hearing aid specialist you trust who will guide you through the evaluation and purchase of a hearing aid. Ask your specialist if he or she accepts incentives to sell a particular device in volume (such as free travel or cash bonuses). It is usually best to find a specialist who has access to several manufacturers in order to meet your needs. Expect to be charged $4,000 for a pair of mid-level hearing aids to $6,000 for upper-end hearing aids.

Most health insurance, including Medicare, does not cover hearing aids. Depending on your state of residence, Medicaid may cover the cost of hearing aids for children. In the authors' experience, many hearing aid insurance policies are disappointing because they cover only low-quality aids. Beware of bait-and-switch insurance schemes.

Most hearing aid specialists recommend purchasing new hearing aids every four or five years. Hearing aids are sold with a one- to two-month money-back guarantee because there is no way to determine in advance whether a hearing aid will help until you try it. The purchase price (less certain fees) is returned if the aid is not acceptable to you.

Economists have long questioned why only 10 to 20 percent of people who need hearing aids actually use them. There are many reasons, but it seems to boil down to the expense. Should a pair of mid-level hearing aids cost eight times as much as a smartphone?

There is now a less expensive option. It is possible to purchase good-quality hearing aids online, direct from the manufacturer to the patient. With a simple computer tablet and interface anyone who can operate a smartphone can program his or her hearing aids. A hearing aid professional is also available at no cost online or by telephone. Hearing aids sold this way range in price from $1,000 to $1,300. Besides cost savings, this alternate sales model provides up to three years of loss and damage insurance and free batteries. There is also the advantage of not having to travel to or wait in the hearing aid office and being able to re-program the aids as often as you wish. On the other hand, buying your hearing aids from a licensed professional, especially a doctor of audiology (AuD), may be well worth the additional expense in terms of professionalism and service.

Assistive Listening Devices

While hearing aids are useful in many situations, some people have difficulties only in very specific settings—like lecture rooms and board meetings, or while watching movies or television.

Assistive listening devices available at theaters pick up sound waves directly from the stage and transmit them wirelessly to the ear device. Televisions have audio output ports that work with wireless headphones. Using a similar strategy, a school teacher, clergy member, passenger in the back seat of the car, or lecturer can wear a tiny microphone clipped to his or her shirt that sends wireless signals directly to an assistive listening device, bypassing the room or car noise as well as making the voice louder.

The use of wireless technology for better signal-to-noise ratio has been around for a long time, but is now being more widely used. Bluetooth is also available for streaming from smartphones.

A sophisticated, relatively new pre-processing strategy aimed at improving the input signal uses a library of acoustic scenes taken from actual recordings of the different kinds of noise in theaters, classrooms, bars, restaurants, etc. The chip recognizes the acoustic scene you are experiencing and automatically switches the hearing aid listening program to the one most effective for that setting. Devices with this capability also perform data logging, keeping a record of how much you use the instrument and under what conditions. This allows a hearing aid specialist to make subtle changes in your device and in counseling you.

Surgical Hearing Aids (Semi-Implantable Hearing Devices)

Bone-Anchored Hearing Aids

Bone-anchored hearing aids (BAHA) are attached to the skull with a screw and transmit vibrations into the bone. Those vibrations then travel through the skull bone to the ear (see the discussion of bone conduction in chapter 1). They are only used for specific types of hearing loss and consist of two parts. First, there is an internal part, basically titanium screw(s) placed into the skull. Some screws actually pass through the skin, allowing the vibrator to attach directly to them. In newer types, the screw does not pass through the skin, but has a magnet just under the skin to hold the vibrator in place. Second, there is an external vibrator that is attached to the screw that is sticking out through the skin (old type) or by a magnet (new type).

Unfortunately, the outer parts of these devices are larger than most standard hearing aids, but they can often be concealed by the hair. When the outer part is removed, a bump is visible under the scalp and sometimes the tip of the screw as well.

Complications are common. With the older technology, up to 38 percent of recipients develop skin infections, 18 percent of the devices fail to attach to the bone, and a second operation is necessary in 20 to 38 percent of adults and 44 percent of children. The implant is rejected in up to 17 percent of adults and 25 percent of children. Because the surgery to attach BAHAs usually requires general anesthesia (posing additional risk) and users experience frequent skin complications, their use is limited to people who cannot benefit from standard hearing aids. Newer technology that eliminates a metal attachment that passes through the scalp should reduce complications.

Middle Ear Implants

Surgically implantable middle ear devices work by attaching a vibrator directly to the bones of hearing instead of to the skull. Middle ear implants are used for conductive or sensorineural hearing loss and bypass the outer ear. Most are partially implanted (battery and microphone remain on the outside). Totally implanted middle ear devices are invisible except for a bump under the skin and allow the patient to hear even while swimming.

However, when these devices are turned off, hearing is actually worse than it was before surgery. The implanted battery may last about four to five years and requires minor surgery to be replaced.

In a study of totally implantable devices, failure rates of 40 percent were reported in a multicenter trial. And patients who said the implant worked reported that their hearing was similar to that attained by using a standard hearing aid. In a more recent study, about half of patients heard better than with a standard hearing aid.

The field of implantable hearing aids is progressing rapidly but still has a long way to go. Your ear specialist can best advise you regarding new developments.

9 | All about Cochlear Implants

■ When Roberto's parents learned that he was born deaf, they knew that they had much to learn about the best way to raise their son. In the end, they decided to follow their county school system's advice: use hearing aids along with signed language. But when Roberto was eight and couldn't read, speak, or communicate well with sign language, they decided to check with a local cochlear implant team. Roberto's parents found that it was too late to receive normal benefit with a cochlear implant. In order to hear and speak normally, he would have needed to be implanted by one to two years of age. Now Roberto was stuck in the middle; he was considered an outsider by the Deaf kids and would never catch up with the hearing or cochlear implant kids. Talk about angry parents . . . ■

Cochlear implants (CI) are computer-based devices that provide hearing to children and adults with nerve deafness. By changing sound waves into tiny electrical signals that stimulate the sense of hearing in the brain cochlear implants replace the function of the cochlea.

How Cochlear Implants Work

In people with normal hearing, the hair cells of the cochlea convert sound waves to electrical/nerve signals that travel the auditory nerves to the brain. But in people who have moderate to profound nerve deafness, many or most hair cells don't work, and the hearing process is blocked.

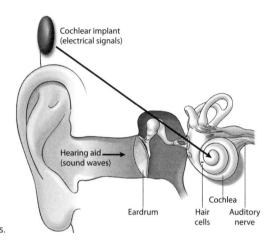

Figure 9.1
Hearing aids make sound waves louder and send them to the hair cells in the cochlea. Cochlear implants turn sound waves into electrical signals and send them on the nerve path to the brain, bypassing hair cells.

Cochlear implants bypass the hair cells. The electrical signals are delivered directly to the auditory nerves.

How Does It Differ from a Hearing Aid?

Hearing aids make sounds louder but if the hair cells are not working (as in nerve deafness) no matter how loud speech is, it cannot be understood. In that case, cochlear implants are recommended because a cochlear implant does not stimulate using sound, just tiny electrical signals.

If a person has enough remaining hair cells to hear adequately with a hearing aid, then a hearing aid is recommended. If not, then a cochlear implant is indicated. It is usually necessary to try a high-power hearing aid for a period of months to see if it can deliver speech understanding before considering a cochlear implant.

Inner and Outer Components

There are two parts of a cochlear implant. The outer part looks somewhat like a hearing aid and is completely removable. The inner part is implanted entirely under the scalp. The two parts communicate through radio frequency waves. Experimental use of cochlear implants that are completely implanted (have no outside part) have so far produced inadequate performance.

External part of the cochlear implant. The external part of a CI contains a microphone, speech processor, antenna, and battery. The microphone changes sound waves to electrical signals. The speech processor (a tiny, programmable computer), processes the electrical waves into signals that are coded for speech understanding.

Both power and the coded signals for hearing are transmitted by radio frequency waves from the external antenna to a matching internal antenna. The two antennas are held in place by a magnet that will last for more than a lifetime. Some external devices are waterproof to allow users to hear while swimming or playing in a pool. Other external devices are not waterproof and must be removed and placed in plastic bags to protect them when used near water.

Internal part of the cochlear implant. The internal part of the CI consists of the antenna, internal microprocessor, a magnet to hold the external antenna in position, and the delicate electrode array that is placed into the cochlea. The internal microprocessor receives radio frequency signals and sends them on to the electrodes. They are placed under the scalp just above and behind the outer ear. There are no internal batteries; power is transmitted from the outside.

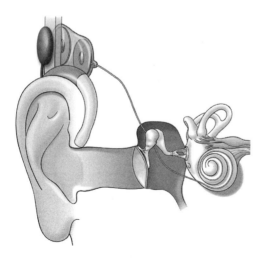

Figure 9.2
This is an image of the right ear as if the head were facing toward you. The external part of the cochlear implant looks like a hearing aid with a transmitting antenna seen overlying the receiving implant that is seen just under the scalp. Just under the intact scalp, the inner part of the implant is seen with the electrode passing through the middle ear into the cochlea.

Candidates for Cochlear Implantation

About 400,000 people have received cochlear implants worldwide, including more than 200,000 Americans. But cochlear implantation is not for everyone. Many people have too much hearing left to qualify and others prefer to live without hearing. Still others may have anatomical or general health conditions that prevent implantation. Communication through signed language (American Sign Language in the United States) is an important alternative to hearing and speaking with a cochlear implant.

The decision to undergo cochlear implantation is made by the patient or family and the cochlear implant surgeon. The surgeon receives input from a team of professionals including audiologists, radiologists, language therapists, psychologists, and others. The family may also receive input from family, friends, other patients, or parents of children who have received implants, and members of the Deaf community.

The cochlear implant team considers type and degree of hearing loss, duration of hearing loss, age of the patient, success or failure of a hearing aid trial, anatomy, health conditions, and commitment to aural communication. Their recommendation is meant to serve the best interest of the patient.

Children

The Food and Drug Administration (FDA) recommends implanting a CI in children born deaf no earlier than twelve months of age. In clinical practice, however, implantation is often performed as early as six months of age. Earlier implantation takes advantage of this important period in a baby's life when he or she normally develops language. Your CI surgeon will consider the child's general health and size when making a recommendation as to the age at which to have the surgery.

In children who are born deaf, implantation after the age of eighteen months results in poorer outcomes in terms of hearing, comprehending language and music, and speaking. After the age of five years, implants may fail to provide hearing that is adequate for spoken language. However, in children who had hearing at birth and lost it later in childhood, implantation anytime in childhood can be very effective.

Adults and Adolescents

Most adult candidates for CI have had hearing and language prior to being deafened. This is called *post-lingual deafness*, and results are generally good. However, if an adult has been severely to profoundly deaf from birth and has communicated primarily in manual (signed) language, the success rate for implantation is lower. This may be because the language areas (temporal and frontal lobes) of the brain may be taken over by visual areas in congenitally deaf signers.

There is no upper age limit on the use of cochlear implants. Many people now lead active vigorous lives into their eighth and ninth decades. Loss of hearing in older adults is associated with social isolation, psychological involution, and dementia. Cochlear implantation is routinely performed for healthy people in their eighties and occasionally in people who are older than ninety. Being able to speak with family members by telephone and to communicate with friends and workers at care facilities is important at this stage of life.

Adolescents who have been deaf from birth frequently have poor cochlear implant outcomes. This may be in part a result of psychological and social pressures. An adolescent may be uncertain of his or her identity as a deaf or hearing person and may be ostracized by friends if the decision is to go ahead with the implant.

Degree of Hearing Loss

Guidelines for implantation have always been prescriptive for the individual patient (that is, one size does *not* fit all) and continue to be modified based on technological advances. Your cochlear implant team will be able to tell you the most current guidelines.

The criteria for implanting adults once required profound hearing loss (90 dB) in both ears. In contrast, criteria are now more liberal and have expanded to include moderate hearing loss (40 dB) in the low frequencies. Figure 9.3 shows the criteria changes in the level of hearing loss over the past three decades.

However, speech understanding, not threshold (see chapter 5), is the main criterion for adult implantation. Speech understanding results using

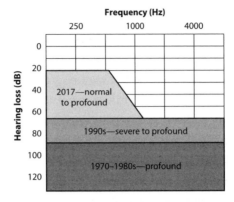

Figure 9.3. Changing criteria for using cochlear implants in adults, 1970s–2017. The amount of hearing loss is shown in the left column. The frequency of the hearing loss is noted at the top. In the 1970s and 1980s, candidates were required to have profound hearing loss in order to qualify for a cochlear implant (bottom box). In the 1990s, people with severe hearing loss in the low and middle frequencies were included (middle box). In 2017, people with normal hearing loss in the lowest frequencies (top box) were added (with emphasis on speech recognition rather than threshold).

hearing aids should be less than 60 percent in the ear to be operated on. Nonetheless, Medicare places further limits on care for seniors and will not routinely cover implantation unless speech understanding is 40 percent or less.

Pediatric criteria are more conservative. Current FDA recommendations include profound hearing loss (90 dB) for children under twenty-four months and severe hearing loss (70 dB) for children over twenty-four months as well as evidence of limited benefit from a hearing aid trial. Threshold is used, rather than speech understanding, for most children with deafness because they do not have adequate language to be tested for speech understanding.

Other Considerations

There should be a strong commitment to living in the hearing world with the enthusiastic support of family and friends. The patient, if the patient is old enough, and his or her family should understand the risks, advantages, and alternatives and have realistic expectations about the probable benefit of the cochlear implant.

Not So Fast

If the cochlea or cochlear nerve is absent, it is not possible to have a cochlear implant. If your child is five years old or more and was born deaf, cochlear implant hearing outcomes are generally not as good as outcomes of younger children. This is especially true if the child communicates in signed language or has no language at all. Poorer results are also common in adolescents, if the cochlea or auditory nerves are not normal or with diminished memory or cognitive function.

Cochlear Implants and Signed Language

By the age of five years, a normal hearing child will have a vocabulary of around 5,000 to 26,000 words. A deaf signing five-year-old will have a vocabulary of only about 200 spoken and/or signed words.

Unfortunately, American Sign Language (ASL) does not have a written form. This is one reason that deaf children who communicate in ASL and attend state residential schools for the Deaf graduate from high school at an average age of twenty+ years with a reading level of a third or fourth grader. Because so many jobs are now in technology sectors and require a high level of literacy, many young Deaf adults may not be able to compete and are often under-employed.

ASL provides many opportunities for communication, education, and socialization. It is at the core of Deaf culture and is a viable alternative to cochlear implantation. Deaf children born to Deaf parents (about 5 percent of all children who are born deaf) do especially well with ASL because they are immediately integrated as members of Deaf culture and are often educated at state schools for the deaf. Cochlear implant teams recognize the value of ASL and respect members of Deaf culture.

However 90 to 95 percent of deaf children are born to at least one parent with hearing. As hearing parents become aware of low reading levels associated with ASL, limited career prospects, and the reality that fewer than 2 percent of neighbors, plumbers, mail carriers, and store clerks will be able to sign with their child, many choose to provide the child with hearing. Fortunately, the Deaf culture is beginning to accept children with cochlear implants as examples of another way to be deaf within their diverse community.

Meningitis

Meningitis is a serious infection of the tissues that surround the brain. It is thought to be more common in children who receive cochlear implants based on studies by the Centers for Disease Control and Prevention (CDC). However, more recent research in the United Kingdom has shown that the risk of meningitis for children who are deaf who have cochlear implants is about the same as children who are deaf who do not have CI. The most common cause of meningitis in implantees is the bacterium *Streptococcus pneumoniae* (pneumococcus). Vaccination reduces the chances of meningitis but does not prevent it.

All children should now be vaccinated against meningitis-causing organisms as part of the normal healthy child immunization program (see table 9.1). Your pediatrician will be certain that all immunizations are up to date. Children are protected if they are up to date on vaccination, even if they will require further vaccination in the near future. Adults receiving cochlear implants are also immunized.

Meningitis is rare but life threatening, and successful treatment depends on early diagnosis. The signs and symptoms to be aware of include

Table 9.1
Recommended Meningitis Vaccinations for All Children and for Adults with Cochlear Implants (CI)

Vaccine	Type of Bacteria	Age	Recommended For
Prevnar-13	*Pneumococcus*	1	all children, adults with CI
Pneumovax	*Pneumococcus*	2–5 Repeat at 65	all children, adults with CI
Hib	*Hemophilus influenzae*	1	all children, adults with CI
Menactra	*Meningococcus*	1	deaf children

Summarized from fact sheet available at http://www.cdc.gov/vaccines/vpd-vac/mening/cochlear /dis-cochlear-gen.htm.

fever, headache, stiff neck, pain caused by bright lights, nausea, and vomiting. Early diagnosis in infants is more difficult because only irritability, sleepiness, and appetite loss may be present.

Cochlear Implant Surgery

Cochlear implantation is usually performed under general anesthesia as an outpatient. It usually requires one to two hours, not including preparation, administration of anesthesia, and time in the recovery room. Current techniques require only a small shave and an incision less than 2 inches long. Preventative antibiotics and a dose of intravenous steroids are often used to avoid infection and reduce inflammation in the cochlea.

After the incision is made, a pocket (or seat) is created under the scalp to securely hold the implanted portion. Various types of seats for the implant are sometimes drilled into the skull. The bone over the mastoid cavity is removed; the facial nerve is located and preserved under a layer of bone. The middle ear is then entered, an opening made in the cochlea, and the electrode inserted into the cochlea. The implantable cochlear stimulator is secured in the pocket and may be held in place by a permanent suture. The area is rinsed with saline, the incision closed, and a light pressure dressing is applied. The device is not activated for a period of days up to a month in order to allow healing.

Outcomes

Cochlear implants improve hearing in over 90 percent of the people who receive them. That is not to say the hearing becomes normal, but the average person understands over 80 percent of what is said in quiet surroundings (normal is 90 to 100 percent).

As noted above, certain patients will have less successful outcomes due to underlying conditions. Children with abnormal anatomy, multiple developmental disabilities, and those over five years of age at the time of the surgery who were born deaf have generally poorer results. Adults who have been deaf for more than twenty years and seniors with memory loss or early dementia should also have lower expectations.

Complications

Some cochlear implant complications are common to any operation, including the risks related to anesthesia and the possibility of infection. Specific cochlear implant complications include slippage of the device, taste disturbance, dizziness, facial nerve injury, and tinnitus.

Infections of the implant are uncommon. Any evidence of redness, swelling, tenderness to touch, or pain can be signs of infection. See your surgeon immediately. Rejection of the cochlear implant is rare (cochlear implants are made of the same biocompatible materials as pacemakers, heart valves, and other implanted devices). The symptoms of rejection can be very similar to those of infection, and sometimes cannot be distinguished without electron microscopic analysis of a removed device.

Rarely, the implanted part of a cochlear implant will migrate out of position. Minor degrees of implant migration do not usually require revision surgery. However, if the electrode begins to slip out of the cochlea, it will probably need to be replaced and re-secured.

Taste disturbance occurs when the nerve of taste obstructs access to the cochlea (about 10 percent of the time) and must be cut. Over a period of months the other taste nerves usually take over, restoring taste.

Dizziness or imbalance problems following implantation are rare and are more often seen in the elderly. If balance symptoms do occur after cochlear implant, they usually last only a few hours or days, but in some cases they may persist for longer periods of time.

This operation requires working adjacent to the facial nerve, which puts the nerve at some risk. Reversible injury occurs in about 2 percent of cases and results in weakness or paralysis of one side of the face. Unless the nerve is severely damaged, it will return to normal function within a month or so. Severe injury to the facial nerve during cochlear implantation is uncommon and may require surgical repair.

Tinnitus may become worse after cochlear implantation in 10 to 15 percent of patients, but it is improved in about 50 to 92 percent of patients who experienced tinnitus before implantation.

10 | Dizziness and Vertigo

■ One night, a week after the car accident, Aisha rolled over onto her right side and her world began to spin. She panicked—stroke, brain injury, heart attack?—and called 911. In the ER they did every test known to man, but the tests were normal. After two days of hospital observation, she went home but continued to spin every time she rolled over onto her right side. A week later, in the ENT office, the doctor had her roll over onto her right side while he watched her eyes for abnormal movement. And that told the story: benign positional vertigo. The treatment was also simple, no prescriptions or surgery, he just maneuvered her head into different positions and the spinning was gone. ■

We would tip over and fall if we relied only on the body's five senses—we also need the forgotten sixth sense: balance. The sense of balance makes it possible for us to stand up on two legs, to walk, and to run. It is a sense we need to live an active life, a sense we take for granted until it fails.

When the balance system is not working, we lose the ability to stand up and move around normally. We might feel a sense of falling, tipping, turning, floating, spinning, or lightheadedness. To make matters worse, problems with the balance system are often associated with secondary symptoms such as blurred vision, nausea, vomiting, anxiety, and sweating.

Disorders of the balance system cause three types of symptoms: vertigo, imbalance, and dizziness. *Vertigo* is a very specific symptom—it is the sensation of motion when you are still. The world may seem to move around you, or you may feel like you are moving. Usually a sensation of spinning, it also can be a sense of rocking, falling, shifting backwards, or any

Table 10.1
Symptoms of Balance System Disorders

Symptoms	Sensation	Caused by Inner Ear?
Vertigo	Spinning, motion	Probably
Imbalance	Unsteadiness, inability to walk a line	Maybe
Dizziness	Lightheaded, tipsy, fuzzy	Maybe
Unconsciousness	Completely out	No

other feeling of actual motion. True vertigo is most often associated with balance system disorders. In the medical sense, vertigo is not the queasy feeling associated with looking down from high places, lightheadedness, tipsiness, or any other feeling that does not create a sense of movement.

Imbalance is the loss of equilibrium, the inability to stand solidly, the inability to walk a straight line, a tendency to fall. *Dizziness* is a less-specific term. It includes sensations of lightheadedness, tipsiness, the feeling you are about to pass out, fuzzy-headedness, difficulty concentrating, and disorientation. Many times, those symptoms of dizziness are not associated with inner ear disorders.

Most balance disorders arise from the ear because the organ of balance is located in the inner ear. The eyes, position sense in the joints, and the sense of touch can also play important roles secondary to the inner ear. The eyes provide awareness of the horizon, which is one reason many falls occur in the dark. Our joints have position receptors that provide awareness of body position. With eyes closed, you can tell if your finger is bent or straight. The sense of touch also plays a role in sensing direction change when sitting, hence the notion of flying by the seat of our pants. The parts of the combined balance system (inner ear, eye, joints, and touch) all come together in the brain.

How the Inner Ear Works

Each inner ear contains three semicircular canals that sense movement (see chapter 1). The canals contain inner ear fluid and motion-sensing

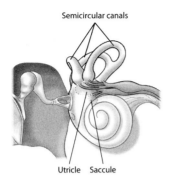

Semicircular canals

Utricle   Saccule

Figure 10.1
Inside the balance system are fluid and hair cells that are bent by head movement (semicircular canals) and gravity (utricle and saccule).

hair cells. When the head moves, inner ear fluid is put in motion, bending the hair cells. Depending on which cells are stimulated, the signals they produce indicate the direction and speed of head movement.

The other part of the inner ear balance system, the utricle and saccule, senses gravity. These gravity receptors contain hair cells too, but also contain tiny calcium crystals. The crystals are heavier than inner ear fluid so they are affected by gravity. When you change head position, gravity pulls on the crystals and bends the hair cells. Note the difference: in the semicircular canals, movement triggers the signals, but in the utricle and saccule, gravity triggers the signals.

Balance System Disorders and Treatment

Before diagnosing a disorder of the balance system, your doctor will make sure that you don't have a serious health condition that is causing the problem. Severe high blood pressure, stroke, heart problems, out-of-control diabetes, electrolyte or hormonal imbalance, low blood sugar, reaction to a medication, seizure, or fainting disorders should all be ruled out or managed first.

During a first, severe episode of true vertigo, you may find the paramedics around you and end up in the emergency room. The ER doctor's first steps will be to rule out heart attack and stroke, control your symptoms, and observe you overnight. Other, non-life-threatening conditions will be defined later and the treatment will be determined by specific diagnosis.

Ménière's Disease

Ménière's disease is a common disorder of the inner ear that causes episodes of true spinning vertigo, fluctuating hearing loss, tinnitus, and pressure in the ear. It is estimated to occur in up to 0.5 percent of Americans (about 150,000 people), and its cause is unknown.

The two major symptoms of Ménière's disease are nerve deafness and attacks of vertigo that last minutes to hours. Two minor symptoms may also occur: tinnitus (often similar to the sound you hear if you hold a seashell to your ear) and fullness or pressure in the ear. Two major symptoms or one major and two minor symptoms are needed to be certain of the diagnosis (see chapter 21). Similar symptoms lasting less than one minute or more than one day are rarely Ménière's. It often begins in one ear but over time may start to affect both ears.

The physician needs to rule out underlying causes, such as acoustic neuroma tumors, perilymph fistula (see below), hypothyroidism, and other less-common disorders that can mimic Ménière's disease.

Ménière's disease is treated in three progressive stages. If a satisfactory response is not obtained with stage one treatment, your physician will move on to the more aggressive treatments outlined in the next two stages. Treatment begins with recommendations for diet, exercise, oral medications, and proper sleep and is effective in 80 percent of patients. If this strategy fails, the next step is often a minimally invasive treatment, such as injections into the middle ear. In the rare case of failure of these first two stages, major surgery, such as cutting the balance nerve, may be necessary. See chapter 21 for more information.

Benign Positional Vertigo

Benign paroxysmal positional vertigo (BPV or BPPV) is the most common cause of vertigo and usually resolves on its own within a few days. BPV is a brief wild spinning sensation caused by a change in head position. Most commonly this first occurs in a middle-aged person when rolling over during sleep and is not associated with hearing loss or tinnitus. BPV may also be triggered by tilting the head back or bending over. A spinning sensation typically takes a few seconds to start and will quickly stop after returning the head to its previous position.

BPV occurs if calcium crystals are displaced from their normal position in the utricle and saccule to lodge in a semicircular canal. Displacement of a crystal may occur after a bump to the head, following an episode of long-term bed rest, or for no apparent reason.

The Dix-Hallpike test is used to confirm the diagnosis. In this examination, the head is turned 45 degrees in one direction and the patient is lowered from sitting to flat on his or her back with the chin up 30 degrees. Delayed onset of severe vertigo and nystagmus (rapid eye-shaking movements that accompany vertigo) confirm the diagnosis in the undermost ear. The test is repeated for the other ear. Videos of this test are available online by searching for Dix-Hallpike video.

BPV is treated by repositioning the crystal out of the semicircular canal using the Epley (or similar) maneuver. With the patient sitting on an exam table, the head is turned 45 degrees toward the involved ear and lowered back until the head is hanging a few degrees off the end of the table. While still hanging, the head is then turned in the opposite direction. The patient is then rolled up onto his or her hip and shoulder with the head facing the floor, and then moved into a sitting position. If necessary, the Epley maneuver can be repeated in the doctor's office or at home. Videos demonstrating the Epley maneuver can be seen online by searching for Epley Maneuver video.

Labyrinthitis and Vestibular Neuritis

Vertigo may result from a simple viral infection of the inner ear called *viral labyrinthitis*. Although the initial symptoms can be severe, they usually resolve over several weeks. Viral labyrinthitis involves the balance and hearing parts of the inner ear and results in vertigo as well as hearing loss and tinnitus. Early cases of Ménière's disease are sometimes misdiagnosed as viral labyrinthitis because the symptoms are very similar. Symptoms are usually restricted to one ear and respond to vestibular suppressants and oral steroids. Antiviral medications in the acyclovir family have been used to treat this disorder but evidence supporting success is lacking.

People with viral labyrinthitis usually have only a few episodes of vertigo over a month or two and make a full recovery. However, in a small number of people, episodes may recur several times in the first year. Recurrence of symptoms is thought to be a reactivation of the virus and subsequent episodes tend to be mild. Some people also develop a chronic phase that is characterized by longer-term imbalance, disorientation, and troubled concentration. This may require further testing of hearing and balance to confirm. The most effective form of treatment for the chronic phase has been vestibular rehabilitation therapy usually consisting of balance exercises provided by a physical therapist.

Unlike the common viral labyrinthitis, *bacterial labyrinthitis* is rare and may be life threatening. Because inner ear fluids are in contact with the spinal fluid, infection of inner ear fluid in bacterial labyrinthitis can lead to meningitis. The symptoms of vertigo and hearing loss along with ear pain, head pain, neck stiffness, and fever are considered to be an emergency and require immediate medical attention. Diagnosis often requires a CT scan and spinal tap and treatment is with intravenous broad-spectrum antibiotics. Bacterial labyrinthitis is a rare complication of otitis media (middle ear infection) and usually results in deafness.

The term *vestibular neuritis* is used when the balance nerve is affected, resulting in vertigo but not hearing loss. This disorder is thought to be viral in nature and has been called *epidemic vertigo* when many cases arise at the same time. Treatment is the same as with viral labyrinthitis.

Acoustic Neuroma and Meningioma

Acoustic neuromas and meningiomas are benign tumors that may affect the balance and/or hearing nerves. The usual symptoms of acoustic neuroma are slowly progressive hearing loss, tinnitus, and imbalance. Symptoms are caused by pressure on the hearing and balance nerves by the growing tumor.

When a hearing test shows significant difference in hearing level or word understanding between the two ears, an MRI with contrast enhancement is performed. The tumors will show up as a small bright mass on the nerves between the inner ear and brain. Treatment depends on the age of the patient, the size of the tumor, and symptoms. Some acoustic neuromas can be monitored with an annual MRI while others may require treatment with radiation or surgery. Your ear specialist will discuss the pros and cons of each option and what might be best for you.

Perilymph Fistula

Perilymph fistula (PLF) is a defect or tear in one of the small, thin membranes that separate the air-filled middle ear from the fluid-filled space of the inner ear. The opening allows fluid to leak into the middle ear air space. The leakage of this fluid may result in vertigo, hearing loss, tinnitus, and a sense of fullness.

Most PLFs are caused by trauma, often including pressure trauma associated with scuba diving. People with congenital malformations of the inner ear are also at risk for PLF. Unfortunately, accurately diagnosing PLF requires surgically exploring the ears of someone who has specific symptoms (vertigo and hearing loss, a history of head trauma or scuba diving, inner ear deformity, and failure of a trial of bed rest).

Bed rest for several days has been shown to be effective in treating most traumatic perilymph fistulas. If bed rest fails, surgical repair of the fistula is possible. This minor procedure takes about 30 minutes and is undertaken under local or general anesthesia as an outpatient. During the procedure, the eardrum is microscopically lifted out of the way and the normally sealed openings into the inner ear, the round and oval windows, are carefully explored for any defects. Leaks are sealed with tiny grafts of the patient's own tissue.

Migrainous Vertigo

Migraine is a relatively common disorder that may often cause headache, visual flashes, and vertigo. But migraine can also cause vertigo that may occur with or without headache and may last hours to days. When one-sided headache and vertigo occur together, the diagnosis is often migraine disease. *Migrainous vertigo* is usually treated with the same medical treatment used for migraine headaches.

Benign Vertigo in Children

Benign vertigo of childhood is an uncommon disorder of children between the ages of one and eleven. Younger children will often respond to vertigo by sitting down, often leaning against a wall, and becoming very still. Their eyes will usually be kept closed, but an adult may be able to observe *nystagmus* (rapid back-and-forth eye movements) beneath the lids or after opening the eyes. The feeling of movement causes fear and anxiety. This disorder is thought of as a migraine variant by some and treated with appropriate dosage of anti-migrainous medications.

Motion Sickness and Mal de Débarquement

Anyone can get motion sickness, but the amount of motion required to trigger it varies from person to person. Motion sickness is more common in children two to twelve years old and in women. With experience, astronauts and sailors learn to progressively accommodate motion. Desensitization treatment by a physical therapist is also effective for many people.

Vestibular nerve suppressant medications block "vertigo signals" from the inner ear to the brain and may also reduce the feeling of nausea. Dramamine (dimenhydrinate) and Antivert (meclizine) and other antihistamine family drugs are available over the counter. Dramamine is stronger but causes more drowsiness than meclizine. Your doctor may prescribe more effective oral, transdermal patch, or suppository medications.

Preventative options for car sickness include driving the car rather than being a passenger or sitting in the front seat instead of the back. Onboard a boat, lying prone with eyes closed or looking at the horizon can also be helpful.

The Brain—Imbalance and Dizziness

In severe vertigo, stroke may be the first diagnosis to consider, especially if there is loss of consciousness or muscle weakness on one side of the face or body. Inner ear problems do not cause loss of consciousness. Disorders of the brain caused by tumors, atypical migraine disease, drugs, multiple sclerosis, and infection can also cause vertigo.

The brain is also affected by medications and metabolic abnormalities. Drugs such as sedatives, blood pressure medications, opiates, marijuana, and alcohol can cause dizziness or vertigo. Such drugs are dose-related and may have severe effects when combined. The most common metabolic cause of dizziness or vertigo is low blood sugar. Some people are hypoglycemic and can avoid episodes of dizziness by eating frequent, small, protein-rich meals. Diabetics may have trouble controlling blood sugar and can develop dangerously low sugar levels that require immediate treatment.

As noted below, *postural (orthostatic) hypotension* occurs when we stand too quickly, causing blood to flow away from the brain. Other circulatory system disorders that reduce blood flow to the brain may cause the same symptoms. Low cardiac output, abnormal heart rhythm, and blockage of arteries may also reduce blood flow to the brain and cause dizziness.

Your physician may ask if your dizziness started at the same time as another event in your life, such as a death in the family or loss of a job. Anxiety is a common cause of dizziness and may require counseling or pharmaceutical treatment. However, medications to treat anxiety may cause dizziness and habituation. Hyperventilation associated with panic attacks also causes dizziness. If you notice rapid deep breathing and tingling of fingers and lips prior to dizziness, tell your physician. Other nonspecific disorders include low blood sugar, especially if there has been a change in your diabetes treatment or rapid weight loss, ear infection, or dehydration.

The Heart and Blood Vessels—Dizziness

The circulatory system (heart and blood vessels) play a role in some forms of dizziness. High blood pressure, low blood pressure, arrhythmias (abnormal heart rhythm), and obstruction of the arteries may all cause dizziness. Circulatory dizziness is usually a lightheaded feeling, but vertigo may sometimes occur.

The most common cause of dizziness in older adults is standing up too quickly (postural hypotension). When standing up quickly, gravity causes blood to flow away from the brain and into the legs. This flow temporarily reduces circulation in the brain. The body's immediate response is for the heart to pump more blood and the vessels in the legs to slightly constrict, forcing blood back to the brain. But as we get older, the body's normal reaction is slower and less effective. Older adults should get up gradually, going from a lying to a sitting position for a few moments and then standing up while holding on to something solid in case they feel faint.

11 | **Tinnitus** | Ringing in the Ears

■ "I saw my primary care physician four times before he sent me to a neurologist. It only took her one visit to say there was nothing wrong and send me to a psychiatrist. Is that doctor crazy or what? I finally made it to an ear specialist."

When I first saw Ms. Wright, she had dark circles under her eyes and appeared anxious. Her main problem was a continuous high-pitched ringing in both ears that had been bothering her for about four months. Her examination was normal and I ordered a hearing test.

Ms. Wright told her best friend the good news: "The hearing test showed that I have a mild hearing loss caused by noise exposure and when they measured the tinnitus, it was mild too. There are no signs of other ear problems and I am not at risk of going deaf. I was already feeling much better, even before we went over the plan for treatment: no surgery, no prescriptions, high chance of success. Yes!" ■

Tinnitus is the sense of hearing something when there is no related external sound. Nonetheless, tinnitus is real; there are well-known causes for tinnitus and it can be measured. Tinnitus is not a disease but a symptom of a physiologic abnormality.

Tinnitus may begin suddenly or develop slowly over several months. There are many different causes: it can be linked to noise exposure, injuries to the ear or head, diseases of the ear, and general medical conditions. Tinnitus can also be a side effect of some medications (high-dose aspirin, other anti-inflammatory drugs, chemotherapy, and diuretics) or a combination of all these causes. Although it can be linked to hearing loss, about 40 percent of people with tinnitus don't have significant hearing loss.

Tinnitus is reported by about thirty million Americans, but fewer than one in three find it disruptive. Most people with tinnitus notice it only in quiet conditions or when focusing on it, but an estimated 10 percent consider their tinnitus disabling.

Classification

There are two major categories of tinnitus (table 11.1):

- subjective—only the individual can hear it,
- objective—a physician can also hear it.

Subjective Tinnitus

This is the common form of tinnitus and accounts for about 95 percent of cases. It may be described as ringing, a high-pitch tone, buzzing, white noise, roaring like a sea shell, or by many other comparisons to everyday sounds. Only the individual with subjective tinnitus can hear it.

Some people experience tinnitus as a recurring melody, indicating origin in the auditory memory areas of the brain. Rarely, subjective tinnitus can be the first symptom of a serious problem such as a tumor pressing on the hearing nerve. Since these tumors usually occur on only one side, tinnitus in only one ear is a red flag that demands immediate investigation.

Objective Tinnitus

Your ear specialist will be able to hear and measure objective tinnitus. Objective tinnitus is much less common than subjective tinnitus, accounting for only about 5 percent of cases. Objective tinnitus is usually *pulse synchronous* (that is, one sound with each heartbeat). Pulse synchronous tinnitus is also called *vascular tinnitus* because it is generated by noisy blood flow in the arteries or veins of the head and neck. Each rhythmic heartbeat brings turbulence caused by too much blood flow or abnormal blood vessels.

Table 11.1
Characteristics of Subjective and Objective Tinnitus

Subjective Tinnitus

- Constitutes 95 percent of all tinnitus
- Usually requires only a simple medical evaluation to diagnose
- Is heard only by the individual
- Is continuous (that is, does not pulsate)
- Varies in loudness and pitch
- Is described as ringing, hissing, buzzing, white noise, roaring, etc.
- Is caused by abnormal nerve signals
- Is affected by stress and caffeine (loudness increases)
- Red flag: one-sided tinnitus

Objective Tinnitus

- Constitutes 5 percent of all tinnitus
- Requires an in-depth medical evaluation to diagnose
- Is heard by the individual and the doctor
- Pulsates
- Synchronous pulsation indicates a vascular cause
- Typewriter tinnitus indicates neuro-muscular cause
- Ear senses the sound, does not cause the sound
- Red flag: all cases

In vascular tinnitus, the ear itself is normal—just doing its job of hearing—in this case, hearing the sound of turbulent blood flow. However, vascular tinnitus can be a symptom of a serious underlying condition and should always be evaluated thoroughly. For example, it may be caused by narrowing of arteries, growth of vascular tumors, and the presence of vascular malformations like aneurysms.

Rarely, objective tinnitus occurs in bursts that are not synchronous with the pulse. This type of tinnitus may have the clicking sound of a typewriter. Involuntary muscle contractions of the palate or middle ear might be the cause. The term for these abnormal contractions is *myoclonus*. The cause is in the motor nerves of the brain, which send irregular signals to the muscles. This type of tinnitus is usually temporary and is often associated with stress or head trauma. In rare cases it may indicate multiple sclerosis, stroke, a brain stem tumor, or a serious head injury.

Loudness of Tinnitus

The estimated loudness of tinnitus is measured with a hearing test performed in a sound-insulated room. The patient listens to his or her tinnitus while a calibrated external sound is put into the ear. When the external sound is just loud enough to block out tinnitus, the loudness of the tinnitus is estimated to be the same as the loudness of the external sound.

About half the people measured in this way say they rarely notice the tinnitus, and the other half say they are disabled by its loudness. It seems that the actual loudness of their tinnitus does not determine whether people feel that it is ruining their lives. The amount of distress caused by tinnitus is instead related to anxiety and depression. In other words, anxiety and depression do not cause tinnitus but make it seem worse. A tinnitus cycle can develop where anxiety, depression, and loss of sleep make tinnitus more debilitating and the worsening tinnitus makes the anxiety and depression more severe.

What to Expect at the Doctor's Office

When you visit the ear, nose, and throat (ENT) doctor's office, you will have your vital signs taken as usual. The measurement of your blood pressure and pulse may be especially important in making a diagnosis.

History

The first step is always a detailed history. Come prepared with answers to the following questions:

- When did the noise start?
- Can you correlate the onset of the noise with any events in your life—for example, an automobile accident, exposure to noise, use of a new medication, scuba diving, death in the family, divorce, job loss, etc.?

- Has the noise been getting worse?
- What does it sound like (for example, many people compare tinnitus to ringing, whistling, hissing, roaring, white noise, pulsing, or a typewriter)?
- Do you hear it in one ear, both ears, or just somewhere in your head?
- Is it worse at a particular time of day?
- Does it get worse when you are stressed (for example, by final exams or an erratic stock market) or when you use alcohol or caffeine?
- How loud is the sound on a 10-point scale (0 = not heard; 10 = like a plane taking off in your head)?
- What is your state of mind (that is, are you feeling anxious or sad and tired)?
- How would you describe the quality and length of your sleep?
- Do you have any other medical problems?
- What medications do you take?
- Have you or other family members had ear problems in the past (exposure to noise, hearing aids, surgery)?
- Do you have associated symptoms, such as hearing loss, ear pain, or ear stuffiness?

Hint: even if you have filled out a questionnaire, be sure to mention any of the above issues directly to the doctor or assistant who does the intake. Bring notes! Avoid telling your doctor a long story about Aunt Rose's tinnitus, because you want your doctor to focus on *you* and on the problem you are having right now.

Physical Examination

The examination will focus on the head, neck, brain, and cardiovascular system. Your ENT will listen for pulsatile tinnitus using a stethoscope over the ear canal, major blood vessels of the head and neck, and the heart. He or she may put pressure on the blood vessels in your neck or ask you to turn your head and say whether these maneuvers change the loudness of your tinnitus.

Hearing Tests

Your hearing will be tested during your first visit. This test can provide important diagnostic information even if you do not suffer from hearing loss. Advanced testing can also detect pulsations of the eardrum.

Specific tests for tinnitus include pitch matching to determine the frequency of your tinnitus, loudness matching to determine loudness of your tinnitus, and a test for *residual inhibition*. In the residual inhibition test, you will listen to a tone that is a little louder than your tinnitus for one minute. When the external tone is turned off, you will be asked if your tinnitus has become softer or gone away and to indicate when it returns to its usual state.

MRI and CT

Imaging tests may be necessary for diagnosis of both subjective and objective tinnitus. Most people with subjective tinnitus will not require any imaging tests, but if your ENT is concerned about the possibility of a benign tumor, he or she may order an MRI.

For those who have objective tinnitus, a CT scan or MRI is often necessary to rule out a benign middle ear tumor, vascular malformation, or stroke. In rare instances, an angiogram (injection of a radio-opaque substance into a blood vessel) must be performed.

Treatment for Subjective Tinnitus

The realistic goal of treatment of chronic subjective tinnitus is management rather than cure. That is, the hope is to reduce the severity and allow the patient to understand and feel in control of his or her tinnitus rather than to feel oppressed and controlled by it. Tinnitus treatment experts have a substantial number of interventions to utilize and management programs are designed specifically for each patient.

The Basics

Most patients find relief after a single visit without referral to a team of specialists or the use exotic equipment. The visit addresses three areas:

Education

Your doctor must take the time to explain what tinnitus is, what causes it, and how working together you will be able to reduce its impact on your life. He or she will review the results of your evaluation to show conclusively that you are safe. If additional tests are indicated to be certain that there is no serious underlying cause of tinnitus these will be explained. Remember, 95 percent of tinnitus is nothing more dangerous than a misperception of sound.

Diet and Exercise

Caffeine, alcohol, and smoking make tinnitus worse. If you drink more than two cups of coffee per day, you will need to gradually switch over to decaf. Also be aware that coffee is not the only caffeine culprit. Many types of tea, soda, and chocolate also contain caffeine. The best ways to achieve healthy sleep and reduce anxiety are to eliminate caffeine and to exercise. A brisk thirty-minute walk each day is a good start. It can also provide a sense of accomplishment and control.

Sleep

Sleep is important for your overall health. If you have a thick neck, snore loudly, or weigh more than you should, you may need to shed some pounds. Your ENT might suggest a sleep study to evaluate your nighttime breathing. Remedies include the use of a CPAP machine or, rarely, surgery. In addition to introducing some exercise into your life, it can be helpful to avoid napping during the day and to establish a routine bedtime.

Unfortunately, use of sleeping pills for a chronic condition may become habituating. You may get less benefit over time and require increasing dosage. Often, sleeping pills leave you feeling hung over and depressed. An alternative is the antidepressant trazodone, which causes drowsiness but is not habituating. The dose of trazodone to treat tinnitus and improve sleep (50 to 100 mg) is usually far less than the dose needed to treat depression (up to 400 mg).

Sound Therapy

Sound therapy uses external sources of sound to reduce the severity of tinnitus. By introducing a louder sound, the softer sound of the tinnitus can be blocked out. It is a simple, time-tested, fast-acting intervention that can be inexpensive and has no side effects. By allowing you to have some control over your tinnitus this method reduces frustration. It can also help a person get used to sound and thus live with the tinnitus more easily. Following are some methods of sound therapy used to promote sleep.

- Everyday background sounds such as music, fans, air conditioners, humidifiers, air filters (or political speeches).
- Generic sound generators can play recordings of birdsong, trains, a babbling brook, or the ocean through headphones or pillow speakers.
- Customized sound generators (also called wearable maskers) look like hearing aids. They are smaller, less conspicuous, customized, and more portable than generic sound generators. They are also more expensive.

Sound therapy is also used when necessary during the day

- Hearing aids amplify what you want to hear and can block out your tinnitus. This is the next step for people with both hearing loss and tinnitus.
- Combination instruments are hearing aids that also generate sound. Standard hearing aids should be tried first, but some people get more tinnitus relief from the additional masking circuit of a combination instrument. Hearing aids and hearing instruments may be very expensive.

Medical Treatment

Medications are not the first line of treatment for tinnitus and although they may be helpful, they should be reserved for cases where the basic methods are not effective alone. It is important to understand that there is no magic pill to cure tinnitus and there are no medications that do not have side effects.

A recent Clinical Practice Guideline from the American Academy of Otolaryngology reviewed various tinnitus studies to identify those that met criteria for evidence-based treatment. Unfortunately, in spite of some promising research, none of the medical treatment studies evaluated by the Guideline met the scientific standards set by its authors. Objections to the studies included poor study design, too few patients in the study, a lack of statistical power, etc. Therefore, the authors recommend against the use of medication to treat tinnitus unless better studies become available. Nonetheless, experienced ear specialists may prescribe medications that do not work for a majority of tinnitus sufferers, but may work for some.

Counseling and Psychotherapy

Because suffering from tinnitus is so intertwined with anxiety, depression, and insomnia, these issues must be dealt with directly when they are severe. There is no stigma associated with seeing a counselor or psychiatrist, and if your tinnitus professional suggests you see one, he or she is not suggesting that you are crazy or that you are imagining the whole thing. Remember, tinnitus can cause anxiety, depression, and insomnia and the circle of tinnitus must be broken somewhere.

Complementary Therapy

Virtually every form of complementary therapy that has been tried for tinnitus has worked on someone. These treatments include massage, chiropractic manipulation, biofeedback, acupuncture, homeopathy, hypnosis, meditation, aromatherapy, etc. When they are successful, they all have common elements: inspiring trust and confidence in the therapist, reducing stress, and improving the sense of well-being. Although these methods have not been shown to be more effective than placebo, many people have felt better and little harm (other than the cost) has been done.

Cochlear Implantation

Tinnitus is not an accepted reason to have a cochlear implant. However, people with tinnitus who need a cochlear implant for deafness have up to a 92 percent chance of tinnitus improvement as a secondary benefit.

Treatment for Objective Tinnitus

Surgery may be recommended in a small minority of cases of objective tinnitus, and in patients with tumors causing subjective tinnitus. Removing the inner ear or cutting the hearing nerve does not improve subjective tinnitus and may make it worse.

Some examples of surgery for vascular tinnitus include correction of severe narrowing of the carotid artery, removal of vascular tumors such as glomus jugulare (see chapter 25), correction of abnormal vascular cross-overs (arterial-venous fistula), and removal of aneurysms (pouch-like bulging of an arterial wall).

Ongoing Tinnitus Research

A 2014 study postulates a two-phase theory of a cause of subjective tinnitus. In the first phase, damage to the cochlea results in abnormal signals being transmitted to the midbrain. As a result, cells of the midbrain adapt by becoming overly excitable. In the second phase, the midbrain cells stop needing signals from the cochlea and fire on their own causing chronic tinnitus.

Another 2014 study tested an approach of sound therapy that won't interfere with hearing and does not require a device to be placed in the ear. In this method, ultra-high frequency sounds (too high for humans to hear) are transmitted to the ear. The result can be a temporary quieting of the tinnitus (*residual inhibition*).

Tinnitus is associated with changes in the brain cells of the auditory cortex that cause them to become hyperactive. Another recent study compares methods that have been used in an effort to reduce hyperactivity of those cells. The most effective treatment was found to be pulsed electrical stimulation of the brain with weak electrical current in a random pattern.

Hyperacusis

Hyperacusis is an increased sensitivity to certain frequencies or volume sounds that are normally not loud enough to cause pain. An estimated 25 to 40 percent of people with hyperacusis also have tinnitus, which indicates that the two conditions may have a common cause. Since mildly loud sounds are uncomfortable for people with hyperacusis, some wear noise protection in normal noise environments. This is thought to lead to further hypersensitivity. Following an evaluation that is similar to a tinnitus evaluation, therapists may prescribe sound therapy to treat hyperacusis.

III | Disorders of the Outer Ear

12 | Swimmer's Ear

■ We could finally afford a swimming pool in the backyard. All summer the family was together, and we had barbecues and invited the neighbors over. But now every time the kids get in the water, their ears start to itch. When they begin to hurt and drain green goop, it means another trip to the family doctor, more ear drops, and no swimming for a week. ■

Swimmer's ear is the usual term for *acute otitis externa,* an infection of the outer ear canal. External otitis affects approximately four people in every thousand each year. Not surprisingly, 80 percent of these infections occur in the summer. Factors that play a role in external otitis besides swimming include living in a humid environment, frequent use of a hot tub, narrowing of the ear canal by wax impaction or bony growth, and use of cotton swabs.

After showering or swimming it is common to experience itching in the ears. Unfortunately, the skin of the ear canal is easily injured by using a cotton swab or fingernail to scratch the itch. By removing the protective layer of wax and by damaging the delicate layer of skin lining the ear canal cotton swabs may lead to external otitis.

Anatomy of an Infection

The ear canal is an opening in the skull that allows sound to reach the eardrum. The outer third of the canal is lined with cartilage and covered by skin. The inner two-thirds of the canal is lined with skin growing di-

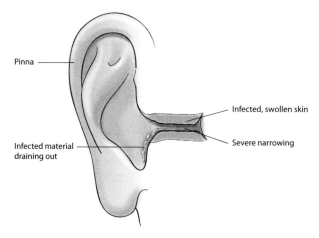

Pinna

Infected, swollen skin

Infected material
draining out

Severe narrowing

Figure 12.1
Clinical appearance of swimmer's ear. Note that the skin is infected and swollen resulting in narrowing of the ear canal. The narrowing can progress to complete obstruction in some cases. Infected material is draining out of the ear canal and can cause a red bumpy rash on the outer skin.

rectly on bone. The skin of the outer third has wax glands that produce the protective coating of the skin called cerumen (see chapter 1).

Swimmer's ear can be caused by bacteria or fungi. Bacteria are normally present on the canal skin, as they are on all skin. In most cases, wax protects the skin from microbes. But when wax has been wiped away by cotton swabs, prolonged water contact with skin of the canal (such as swimming) reduces resistance to bacteria or fungi. The same process occurs when the skin is broken by scratching it.

Bacterial Infection

Bacterial infections make up about 95 percent of acute external otitis in temperate climates. The most common bacteria to cause external otitis are *Pseudomonas* (40 percent) and *Staphylococcus* (25 percent), both of which tend to be resistant to commonly used oral antibiotics such as amoxicillin but sensitive to antibiotic ear drops. The most prominent symptom is ear pain and tenderness. The doctor will look into the ear with an otoscope slowly and carefully, since the infection may cause tenderness. The doctor may need to remove infected debris to see the canal properly.

Fungal Infection

Fungal infections account for less than 5 percent of acute external otitis and are even less common in dry climates. *Candida albicans* is a white yeast that can cause a mild infection, especially in people who wear hearing aids. This is the result of the buildup of moisture under the hearing aid, irritation of the skin of the ear canal, and rubbing away of the wax layer.

Another fungi, *Aspergillus,* causes more serious infections that may be more difficult to eradicate. *Aspergillus* fungus produces creamy, thick, and sticky discharge. Fungal infections can often be identified by the presence of tiny bristles (the fungus itself) that are often black. When the organism causing swimmer's ear is easily identified, a culture may not be necessary if treated early. In cases of resistance to treatment or in advanced cases, a culture is indicated.

Prevention of Acute Otitis Externa

Some children and adults have recurring episodes of otitis externa. The first step in management is to be certain that infections are completely cleared and the ear canal skin and cerumen are back to normal before allowing water to enter the ears. This may require two to six weeks of dry ear precautions. The authors recommend placing a compact cotton ball in the ear canal opening, then covering the outside of the cotton with petroleum jelly to prevent water penetration. This method has the advantage of using disposable and non-damaging cotton. Reusing plastic ear plugs may prolong infections by re-contaminating the ear as well as by damaging fragile skin.

Cotton swabs, fingernails, and all other objects must be kept out of the ears. The smallest bruise or scratch may lead you right back to the doctor's office and the more times infection occurs, the more vulnerable the ear is to re-infection.

Staying out of the pool or the lake may not be practical during a vacation or hot summer weather, but it is the most effective prevention for many people. Keeping most water out of the ears is possible for surface swimmers by using ear plugs.

When using ear plugs, ear drops with acetic acid (white vinegar), alcohol, or a 50/50 combination of acetic acid and rubbing alcohol can be helpful after swimming. These preventative drops help evaporate moisture that got past the plugs while reducing bacterial growth by being slightly acidic. Ear dryers, available in pharmacies and online, are also useful to remove moisture from the ears (see below).

Ear Plugs

Many different types and styles of ear plugs are available. Your doctor will recommend a type that has worked well for his or her patients and your local pharmacy will have a number of brands you can try. The variety is even larger online.

Be sure to get the plugs designed for swimming rather than noise protection. One effective type is a moldable silicone plug that is custom fit at home. One drawback of using ear plugs is that they tend to leak a bit. Try to find a brand that makes a secure seal in your ear. For the most part, over-the-counter ear plugs work as well as the $100 custom plugs at the audiology office. Headbands are also available to keep the plugs from falling out. Be sure to wash the plugs carefully and allow them to dry before re-use.

Drying the Ear Canal after Swimming

If water enters the ear canal it can be partially removed. Place a piece of tissue into the ear opening. It does not need to go in more than 1/10 inch; just breaking the surface tension will cause the water to wick out.

Hair dryers can also be used for this purpose, although they are not especially effective since circulation of air cannot be established in the ear canal. Pull the ear back gently to open the canal, place the dryer on low heat at least 6 inches away for about two minutes.

Specialized ear dryers are also available. Rather than blowing from the outside, these have a safety tip that extends a short distance into the canal. Some blowers have an electrical fan and others use a squeeze bulb to get the air flowing. They are available online as well as in many pharmacies.

An alternate method of drying the ear canal is to use over-the-counter drops that contain a rapidly evaporating liquid like alcohol after swimming.

Table 12.1
Preventing Swimmer's Ear

Method	Description	Timing
Ear plugs	moldable silicone	Before swim
Acidifying drops	2% acetic acid	Before/after swim
Drying drops	vinegar/alcohol	After swim
Air dryers	electric fan or bulb	After swim

Of course these products cannot be used if there is a hole in the eardrum and are usually not adequate to treat an existing infection. A variety of liquid drying agents can also be found online and in pharmacies. Beware of hydrogen peroxide containing drops that are designed for wax removal. These can damage and further reduce the skin's resistance to infection.

Two percent acetic acid (Domboro Otic, Vosol) is used to prevent infections by slightly acidifying the ear canal. The acid environment can keep both bacteria and fungi from getting a foothold. At times, acetic acid drops are also used to treat an established infection. Acetic acid ear drops are available by prescription or over the counter at most pharmacies or online.

A solution of half alcohol and half vinegar is very effective in preventing swimmer's ear and can be mixed at home. Again, be sure to check with your doctor for directions before starting any treatment. In resistant cases, a combination approach may be called for. Place drops in the ear canal, then slowly insert ear plugs before swimming. After swimming, remove plugs, dry the ear with a blower then add more drops.

Table 12.1 summarizes prevention of swimmer's ear.

Treatment of Acute Otitis Externa

Dry Ear Precautions

By keeping the ear totally dry and using prescribed antibiotic ear drops you can usually clear up an infection in the early stages within a few days. Amoxicillin is not indicated for external otitis. Dry ear precautions are critical to getting better—no cheating. If the ear is draining, it is fine to catch the drainage with a loose piece of cotton, but do not block the opening, trapping pus inside. The cotton should be changed when moist.

Absolutely no swimming or water in the ear should be permitted until your doctor clears you. This may be two to six weeks after the infection is eliminated. When showering, protect the ear canal by placing a cotton ball firmly in the opening. The outside (only) of the cotton ball should be coated with petroleum jelly in order to keep moisture out. Petroleum jelly–coated cotton balls are better for this purpose than store-bought ear plugs because they are less traumatic to insert and can be thrown away afterwards, rather than re-used—minimizing the risk of re-infection. When bathing, rather than showering, do not submerge your head in the bath water.

Antibiotic Ear Drops

Antibiotic ear drops are preferred over oral antibiotics. Some antibiotics cannot be taken by mouth or injection, and the concentration of antibiotics in ear drops is much higher than can be achieved by oral antibiotics. However, for advanced bacterial infections that invade the deeper tissues, or for people whose immune systems are weakened, oral or injectable antibiotics may also be necessary.

Some common antibiotic drops include neomycin and polymyxin B (beware, 10 percent of the population may become allergic to neomycin), ciprofloxacin (Cetraxal), ofloxacin (Floxin Otic), and gentamicin. Ear drops that contain antibiotics usually require a prescription. Antibiotic ear drops are often combined with steroid solutions in order to reduce swelling, inflammation, and pain.

When swimmer's ear is caused by a fungus, it is also treated by meticulous cleaning and drops containing acetic acid. Antifungal drops, such as nystatin and clotrimazole, can be effective in treating the fungus *Candida* but are less effective in treating the common *Aspergillus* fungi. Your ear specialist may apply antifungal liquids to the ear canal, such as gentian violet or Cresylate, after cleaning. If topical methods are not effective, oral itraconizole may be prescribed but has a small risk of complications that involve the heart and liver.

Referral to an Ear Specialist

If simple treatment methods are not effective, you will be referred to an ear specialist. External otitis creates infectious debris that can block drops from reaching the bacteria or fungi. Your ear specialist's first step is to culture and remove the pus or debris using an ear microscope. Careful cleaning reduces the number of bacteria or fungi and allows the drops to reach the infected skin surface. Repeat cleaning may be necessary every few days.

In more advanced cases of otitis externa, the skin swells and may even close off the ear canal. This usually occurs when the infection has invaded the tissues under the skin lining and is called cellulitis. Cellulitis is a very painful situation and the swelling will prevent the drops from reaching infected tissue. When the ear canal is swollen shut, your ear specialist may need to insert an expanding wick. The wick will gradually open the canal and bring antibiotic drops to the infected area. Insertion can be momentarily painful but will usually result in quick relief of discomfort. Cellulitis generally requires oral antibiotics as well as antibiotic drops.

Complications

Chronic Dermatitis and Otitis Externa

Outer ear infections may come and go frequently over many years. This is generally caused by an underlying condition of the skin. *Seborrheic dermatitis* is similar to dandruff and often occurs in the scalp, eyebrows, and ear canal. It reduces the ear canal skin's ability to fight infection.

In cases of recurrent otitis externa, the present episode is treated in the usual way (dry ear precautions, antibiotic ear drops, etc.). Once the acute infection is controlled, it is time to treat the underlying skin condition that would otherwise lead to further infections. Seborrheic dermatitis can be treated with oil-based steroid drops.

Psoriatic dermatitis also affects the ear canal and is a more severe but less common disorder than seborrhea. Psoriasis causes a red, flaking rash but with thick silvery scales and plaques that can bleed when removed. It is also common on the elbows and knees. In some cases, psoriasis and otitis externa should be jointly managed by ear and skin specialists.

Chronic itchy ear may also result from the skin condition called *eczema*, which is a red, itchy rash, usually of unknown cause. Dermotic is an oil-based steroid ear drop that may be used in such cases. Sometimes it is necessary to stop all medications to allow the skin time to recover.

Contact dermatitis of the ear canal can also result from the use of hearing aids, hairsprays, and ear drops that contain neomycin. The allergic rash is red and bumpy and appears at the opening of the canal or where the drops overflow onto the pinna. Once recognized, it is usually treated with topical steroids and withdrawal of the offending substance. Hearing aids that irritate the skin can also be replaced by a type made with non-allergenic plastics.

Spread of Infection

Infection of the outer layers of skin can spread to involve the deep layers (*cellulitis*), the cartilage lining (*perichondritis*), the cartilage itself (*chondritis*), or bone (*osteitis*). Culture of the infection is necessary along with prescription use of oral or injectable antibiotics. Quinolone antibiotics are the drugs of choice (for example, ciprofloxacin), having been shown to be safe and effective in children and adults.

When the cartilage lining is infected (perichondritis), the outer ear swells and turns red. The skin becomes very firm. As with cellulitis, the common bacteria found are *pseudomonas* and *staphylococcus* species and the drug of choice is a quinolone antibiotic. If a mushy soft area is present, it may represent an abscess (accumulation of pus) and must be drained. This fluid under the skin can also indicate that the cartilage itself is infected (chondritis). Infected cartilage often must be cut away because blood supply of cartilage is too meager for the antibiotics to be effective. Rarely, a non-infectious inflammatory condition involving cartilage in several areas of the body may mimic infectious chondritis of the outer ear. This may be caused by an autoimmune process and require steroid treatment.

Special Caution: Diabetes and Immune Suppression

Older people with diabetes, especially if it is poorly controlled, are at risk for a very aggressive form of external otitis. The same is true for people who have had organ transplants, HIV, are taking high doses of steroids, or undergoing cancer chemotherapy. All share a difficulty fighting infection because their immune systems are compromised by the underlying illness or treatment.

In people whose immune systems are suppressed (people who are said to be immunosuppressed), a simple bacterial or fungal infection may invade the underlying tissues and begin to spread through the bones of the skull. An earlier term for this condition was malignant otitis externa, but that term has largely been replaced by the terms *necrotizing otitis externa* or *skull base osteomyelitis*.

The diagnosis of skull base osteomyelitis is based on a high degree of suspicion when a patient with a weakened immune system gets an ear infection. Its hallmark symptoms are ear drainage and deep head pain.

Delay in diagnosis or appropriate treatment can lead to severe, chronic skull infection, and fatalities have been reported. The diagnosis requires microscopic cleaning of the ear canal and observation for granulation tissue, a reddish buildup of tiny blood vessels in response to bone infection of the ear canal. CT, MRI, and other imaging studies can be used to confirm the diagnosis and chart the progress of treatment. In many cases, the patient is hospitalized and placed on a combination of long-term intravenous antibiotics. Sometimes, healthy young patients who are identified early may be treated without hospitalization.

13 | Ear Wax and Foreign Bodies

■ It was the end of December. I had overslept and was rushing to brush my teeth, shave, and cotton swab my ears after a shower. Unfortunately, my son chose that moment to bang the door open right into my elbow, which was bent to hold the swab in place. First I felt the clap of pain, then I realized I couldn't hear from my right ear.

Two hours later I was in the ENT office, on my back and under a microscope. The doctor removed the blood from my ear canal with a tiny vacuum. She said there was a small tear of the skin that would heal quickly and that the eardrum was OK. I made two New Year's resolutions: get a louder alarm clock and lose the cotton swabs. ■

Ear Wax Impaction

Approximately 12 million people in the United States experience buildup of ear wax each year that leads to some sort of blockage (impaction). Why does the ear create wax (cerumen)? What factors contribute to impactions? And how best can you prevent and treat cerumen impactions?

The ear canal is about 1 inch long (24 mm). The outer one-third of the canal is made up of cartilage, whereas the inner two-thirds is made of bone. The ear canal is lined with skin, much like the skin that covers the rest of the body, but there are some important differences.

The external ear has a unique self-cleaning mechanism. Like all skin of the body, outer layers of skin cells are lost every day. In the ear, these cells migrate outwards (from eardrum to ear canal opening) carrying debris

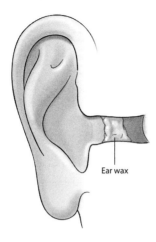

Ear wax

Figure 13.1
The skin of the outer third of the ear canal contains glands that make cerumen (ear wax). Like the skin of your face or hands, layers of ear canal skin shed every day. These layers of skin join glandular secretions to make wax. Like a slow conveyor belt, the wax carries out dirt that may have entered the canal.

with them. They become part of cerumen (ear wax) and extrude on their own from the ear opening. The wax is key to protecting the ear canal skin.

Cerumen glands are similar to both mammary glands and sweat glands. The cerumen itself has several functions. It helps prevent outer ear infection because it:

- protects the skin from water (picture your toes after a long bath; all white, crinkly and easy to bruise, which can set the stage for an infection),
- is slightly acidic, making bacterial growth difficult, and
- contains antimicrobial proteins that attack bacteria.

Together these factors contribute to an antibacterial and antifungal effect of cerumen.

The normal migration of cerumen through the ear canal usually prevents the development of impaction. Use of cotton swabs to assist this natural process has a tendency to push the wax back into the ear canal. Think of a muzzle-loading rifle, where the powder is packed deeply into the barrel by a plunger. You may think you're removing wax because the tip of the swab is stained yellow, but the main effect of most swabbing is to force cerumen deeper in the ear canal. Cotton swabs (or worse, bobby pins or other sharp objects) also cause lacerations of the delicate canal skin, resulting in infection, and occasionally they perforate the eardrum.

Other factors may also lead to cerumen impaction. Some people have narrowing of the ear canal that is present from birth or a result of trauma or infection. A narrow ear canal (smaller than 3 or 4 mm in diameter) may obstruct the exit of wax, causing a buildup. In a similar way, some types of hearing aids may block the opening and lead to cerumen impactions. Finally, patients with dermatologic conditions of the ear such as seborrhea (akin to dandruff) or psoriasis (an autoimmune disease where, among other problems, the skin layers flake off very rapidly), may find that accumulation of dead skin at the opening of the ear canal leads to a blockage.

Symptoms

Cerumen impactions cause a range of symptoms. Most commonly patients experience a feeling of fullness or a clogged sensation in their ear. Itching may also be present. Hearing loss and the sense of hearing your own voice echo in the blocked ear occurs if the ear canal is completely blocked. These complete-blockage symptoms may seem worse after showering, as water in the ear canal causes the wax to swell or fill in a small air gap to form an airtight blockage.

Prevention

People with a tendency to get wax impactions may require special preventative management. Softening agents may help wax come out of the ear canal more easily. Baby oil, mineral oil, and olive oil have all been used effectively. On the other hand, using hydrogen peroxide–based cleaners and detergents to loosen or remove wax can damage the ear canal skin and lead to infection when used repeatedly. In rare cases, regular removal of ear hair or cerumen by a physician is necessary. Even rarer, people with a narrow ear canal may benefit from surgical enlargement.

Treatment

The safest and most effective means for removing impacted cerumen is by an ear doctor with micro-instruments under a microscope. Gentle irrigation, in either a physician's office or using home ear wax removal kits, is appropriate to soften and dislodge wax. Sometimes called "syringing," this process involves directing water toward the back wall or roof of the ear canal in an attempt to wash out the wax plug. The effectiveness of either of these techniques may be enhanced by softening the wax, typically with olive or baby oil for two to three days prior to removal.

Ear candling has been promoted as an alternative medicine technique for removing impacted cerumen, but we do not recommend it. In this technique a hollow candle is placed in the ear and lit. Proponents claim that the heat creates a vacuum in the hollow candle that draws wax out of the ear. As proof, they demonstrate the wax accumulation at the base of the candle. That wax, however, has been shown to be from the melted candle combined with soot from the wick. Candling does not create a vacuum or remove wax. Furthermore it may cause superficial burns when hot candle wax drips down into the ear canal.

Foreign Bodies in the Ear Canal

Any outside material that becomes lodged in the ear canal is a foreign body. It is relatively common, and, not surprisingly, children and adults tend to get different types of objects stuck in their ear canals.

Once children gain manual dexterity (around the toddler stage) they are able to place small objects, such as beads or other small toys, into their ears, where the foreign bodies become lodged.

Adults, on the other hand, may experience insects gaining entry into the ear during sleep. It is also common to find the tips of cotton swabs or components of a hearing aid in the ear (often a tip or "dome" can be left behind when the rest of the aid is removed).

Treatment

Once lodged in the ear canal, foreign bodies should be removed within a day or two with one notable exception (disk batteries). If foreign bodies are left in the ear canal they cause swelling of the skin, obstruction, wax buildup, and infection. Removal of disk batteries requires immediate action because these tiny batteries can destroy the ear canal skin, underlying bone, and the eardrum if not removed quickly.

If living insect(s) are present within the ear canal, they may voluntarily crawl out toward a flashlight in a darkened room. If removal is nec-

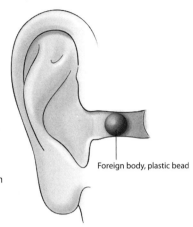

Foreign body, plastic bead

Figure 13.2
Round or hard foreign bodies that are deep in the ear canal usually require using a microscope and micro-instruments for removal.

essary, sometimes irrigating the ear with tap water will flush the insect out without killing it. Nonetheless, it is usually necessary to euthanize the insect before attempting removal. This can be done by filling the ear canal with alcohol or mineral oil.

The best way for a doctor to remove a foreign body depends on its size, shape, and texture. Irregularly shaped items and soft compressible objects can commonly be removed in the emergency room by an ER physician. A smooth, hard object, such as a bead or popcorn kernel, is more likely to require an ENT who will remove it using a microscope and micro-instruments.

One recent study demonstrated a success rate of only 34 percent for removal of smooth rounded objects by non-ear surgeons. If a foreign body is deep in the ear canal, or one attempt is unsuccessful, the patient should be referred to an ear specialist. If the ear canal has been cut or bruised, or if a child has been traumatized in the ER, it may be necessary to remove the foreign body under general anesthesia in an operating room.

14 | Malformations of the Outer Ear

■ Two excited new parents are dismayed to learn that their newborn baby girl has a deformed outer ear. They are concerned about her hearing and need to know what options are available to improve the appearance. ■

Anomalies of the external ear occur in about one in every six thousand newborns. There are many types of malformations that can affect the pinna as well as the ear canal (see chapter 1). The causes of ear deformities may be genetic or a result of exposure to environmental toxins. The term *microtia* means small ear (micro + otia), but your doctor will use it to indicate size as well as abnormal shape and position of the pinna. Severe microtia frequently occurs with *atresia*, complete closure of the ear canal. *Stenosis* refers to narrowing or partial closure of the external auditory canal.

Genetic deformities may involve only the ear or they may occur with other anomalies. Groups of anomalies are called *syndromes*. Common syndromes causing external ear deformity include Goldenhar syndrome (one side of the face is smaller than the other), branchio-oto-renal syndrome (ear, neck, and kidney involvement), Down syndrome, Treacher Collins syndrome (small cheek and jaw bones with eye and ear anomalies), and others.

Non-genetic malformations have been associated with viral infections, malnutrition, low oxygen levels, exposure to drugs, and other causes. The reason for the malformation is often not known.

Microtia

Microtia is more common in boys than girls and usually involves just one ear (the right more than the left). It can be classified into a confusing array of anatomical types. The following is a simple, three-category system that is based on how much of the normal ear structure is present.

- Grade I (mild) microtia: the ear exhibits minimal deformity, such as low-set ears, lop ears, cupped ears, and folded ears. All major structures of the external ear are present.
- Grade II (moderate) microtia: moderate deformity is present. All external ear structures are present but they are small and misshapen.
- Grade III (severe) microtia: (sometimes called peanut ear) is severe and has few or no recognizable structures. The ear lobule is usually present but positioned forward of its typical location.

Treatment

Grade I microtia can sometimes be treated in early childhood with taping or plastic molds that reshape the pinna while it is still moldable after birth. Correction of Grade II microtia requires plastic surgery based on changing the shape of the underlying cartilage.

The most common procedure to reshape moderate abnormalities is called *otoplasty*. Otoplasties create a normal fold of the ear, improving the appearance of the pinna and repositioning it closer to the skull. Frequently used techniques include recreating the antihelix by placing sutures through the cartilage, sometimes with cartilage sculpting. *Hematoma*, an accumulation of blood adjacent to the cartilage, is the most common complication following otoplasty, occurring in approximately 3 percent of all patients.

Grade III microtia repair is much more involved and may require multiple (often four or more) complex procedures. Reconstruction uses cartilage transplanted from the ribs and the results are often not especially attractive. Be sure to see the plastic surgeon's post-op photographs of his last twenty patients and find out how many operations will be necessary before making final decisions.

The following will give you an idea regarding the extent of Grade III microtia repair surgery:

Stage I: Cartilage Implantation. Rib cartilage is typically removed from the lower ribs. The goals of this stage include creating a cartilaginous framework for the new ear that is symmetrical to the normal ear. The rib cartilage is carved to look like a pinna and implanted under a pocket of skin in a position equivalent to the opposite ear. The major complication of this stage is air leakage into the chest cavity that can sometimes (but rarely) result in lung collapse.

Stage II: Lobule Transfer. This procedure is often performed two to three months after Stage I reconstruction and aligns the ear lobe (usually present) with the reconstructed cartilage framework.

Stage III: Post-auricular Skin Grafting. The back of the carved rib cartilage is separated from the skin pocket and skin grafted. This allows it to project from the skull. This step is performed about three months after Stage II reconstruction.

Stage IV: Tragal Reconstruction and Soft Tissue Debulking. The tragus is created with a separate cartilage graft. This is often performed several months after Stage III reconstruction.

Another approach to severe microtia is to use a prosthetic pinna. In general, prosthetics are safer and more normal in appearance and feel. They are custom made and usually covered by insurance. Prosthetic ears are attached by glue or by magnets. The prostheses are removed nightly and washed. If magnets are used, a minor procedure to implant an inside-the-skin bone-anchored magnet is necessary. The inside magnet will hold an outside-the-skin magnet in the base of the prosthesis firmly in place. Prosthetic ears need to be replaced every five years or so, depending on the wear and tear from swimming and contact sports.

Atresia and Stenosis of the External Ear Canal

Congenital anomalies of the external ear canal range from mild *stenosis* (narrowing) to complete *atresia* (total obstruction). These are often seen in association with malformations of the external ear and the structures of

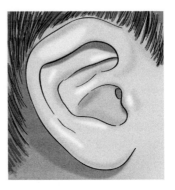

Figure 14.1

Microtia can range from having no pinna at all to having an ear that is small but otherwise normal in appearance. The greater the deformity, the more likely it is to be associated with atresia (total obstruction).

the middle ear. Atresia and stenosis are easily recognized by your pediatrician, but should be confirmed by an ear specialist. Using an ear microscope, the physician will examine the ears and then conduct a hearing test. The typical pattern of hearing loss in affected ears is a moderate conductive hearing loss of 50 to 70 dB. If both ears are affected, this is enough to cause language development delays if not recognized and treated.

The ear specialist will also recommend that the child have an X-ray or CT scan prior to one year of age. These imaging studies are used to identify specific abnormalities, such as those of the middle ear bones and facial nerve, as well as to be certain that no skin cyst is developing as a result of trapped skin left behind during development (see figure 14.1).

Treatment

If conductive hearing loss is present in both ears, it is necessary to apply a bone-conduction hearing aid as soon as possible. Bone conduction hearing aids are vibrators that are held against the head by elastic head bands. They vibrate the skull to stimulate the cochlea directly without passing sound waves through the obstructed ear canal.

Two options for definitive treatment of atresia (total obstruction) and stenosis (partial obstruction) are available. The first is surgical repair, which is usually recommended for stenosis and favorable types of atresia. The second is use of a bone-anchored hearing aid (BAHA). BAHA is basically a skull vibrator on the side of the head. Unlike a bone conduction hearing aid, which is held in place by an elastic band, BAHA requires placement of a bone implant behind the ear that then couples directly or magnetically with a bone-vibrating hearing aid. It is usually recommended in unfavorable types of atresia because it is safer and more effective.

15 | Bony Growths of the Ear Canal

■ Kelly's family doctor found a growth in her ear canal, and now Kelly is in the ear specialist's office to see what's going on. It turned out to be pretty simple: a slow-developing benign bone growth caused by a lot of exposure to cold water. We'll check it again in two years to make sure it is not starting to block the canal. ■

Most bony lesions of the ear canal are caused by swimming in cold water. They are almost always benign growths. They are frequently referred to as *osteomas*, but they are more accurately called *exostoses*. In most cases, neither causes symptoms but they may be discovered during a routine ear examination.

Problems occur if these growths enlarge enough to block the ear canal. This may lead to cerumen impaction and recurrent outer ear infections (if the growth traps cerumen and water). In rare cases, bony lesions may grow into the eardrum leading to conductive hearing loss. Exostoses and osteomas are separate clinical entities that differ in their cause as well as appearance.

Exostoses

Exostoses (the plural form of exostosis) are growths of new bone in the bony portion of the ear canal. They are caused by repeated exposure to cold water, which stimulates the lining of bone to produce new bone layers (see figure 15.1). Surfers, who often find the best waves in colder water, are commonly affected by this condition. In addition to the cold water, surfers are also exposed to winds that cause evaporation of water from the ear canal.

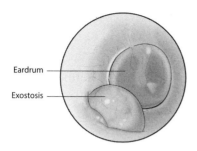

Eardrum

Exostosis

Figure 15.1. Bony growths of the ear canal are almost always benign but may enlarge over time to obstruct the canal. A small proportion require surgical removal.

Evaporation further cools the ear, stimulating bone growth. Exostosis is often called *surfers' ear* in areas where the sport is popular.

One recent study found that exostoses were larger in ears that faced prevailing winds. Up to 73.5 percent of surfers have exostoses. Multiple growths, affecting both ears to different degrees, are frequently found in adults with a history of cold water surfing.

Exostoses have tiny layers of new, reactive bone that is formed on top of the normal bone of the ear canal. When studied under a microscope, these layers can look like the consecutive rings seen on growing trees. The lesions are broad-based and lack a central core.

Osteoma

Osteomas, unlike exostoses, are single, involve only one side, and have a stem-like base. Osteomas are more common in children and young adults. Most of the time osteomas are discovered during routine ear examination. They tend to be whitish, smooth, round masses covered with healthy skin and partially obstructing a view of the eardrum.

Osteomas differ from exostoses in that the normal layers of bone are organized around a central core that contains blood vessels. They attach to the underlying bone by a narrow stem and in some ways resemble a tiny mushroom.

Diagnosis and Treatment

A history and clinical exam are usually sufficient for diagnosing bony lesions of the ear canal. In circumstances in which the diagnosis is in doubt, or to assess the extent of the disease when the ear canal is completely obstructed, CT scans can be useful.

Most bony masses require no treatment and have no associated risk. They should be observed every two to three years to see if further growth begins to block the ear canal. Blockage could lead to recurrent wax impactions or infection and may signal the need for surgical removal of the tumor.

Surgical removal of exostoses is done under a microscope. Skin of the ear canal is removed from the surface of the growth and replaced after the growth is surgically removed. Care is taken to avoid damage to the eardrum in order to preserve hearing. The abnormal bone is removed with a drill and the patient is left with a skin-covered, healthy ear canal.

In smaller exostoses, it is possible to operate through the natural ear opening. This avoids an incision. This approach may not be possible when the growth is large or the canal is small. When more access is necessary, an incision is made behind the ear or in the cartilage of the ear canal. Either incision provides excellent access. These operations have a high rate of success and few complications. However, continued cold-water surfing may result in recurrence of growths.

Surgery for osteomas is usually simple because they rise on a narrow stem. Osteomas can be removed under local or general anesthesia as an outpatient procedure. Recurrence is uncommon.

16 | Cancer of the Outer Ear

■ A fisherman in his early sixties went to see his dermatologist because he has a bleeding lesion on the upper rim of his ear. It has been there for about two years but recently has been getting larger. ■

Non-healing lesions of the ear should be carefully evaluated for features that suggest malignancy. Ulceration, bleeding, odd pigmentation, asymmetry, nodularity, and rapid growth are all features that should prompt concern. It is often necessary to perform a biopsy of these types of lesions in order to determine if there is evidence of cancer. If the biopsy indicates skin cancer, treatment strategies are based upon the size of the tumor as well as the type.

Basal Cell Carcinoma

Basal cell carcinomas are the most common malignant neoplasm of the ear, representing 45 percent of ear cancers. Chronic long-term sun exposure is the predominant cause of basal cell carcinoma. Specifically, ultraviolet B (UVB) radiation has been identified as a major carcinogen. The incidence of cancer increases with age. Other risk factors include fair skin, outdoor occupations, and a history of previous skin cancer.

People may initially develop a skin lesion that is nodular (lumpy), ulcerated, and/or bleeding. Basal cell carcinomas of the ear typically occur on the back surface of the pinna. Imaging may be used to evaluate advanced disease with tumor extension to the adjacent temporal bone and soft tissue structures of the head and neck. The overall rate of basal cell cancer metastasis is extremely low, at less than 1 percent.

Treatments of ear basal cell carcinoma may be either surgical or non-surgical. Non-surgical options include treatment with chemicals to destroy the tumor such as topical 5-fluorouracil. This may be successful with small superficial lesions, but is far less effective with larger lesions. Radiation therapy is also an option, but is usually reserved for patients who are poor surgical candidates because of other conditions or for people whose lesions are too extensive to be surgically removed.

Surgery is far more commonly used to treat basal cell carcinomas affecting the ear or the ear canal. Cryosurgery, which involves freezing lesions with liquid nitrogen, may be used to treat small basal cell carcinomas (< 1 cm) with well-defined borders. Standard surgical excision (removal) is the mainstay of treatment. Ninety-five percent of basal cell carcinomas < 2 cm in size can be successfully treated with local excision with a proper margin (normal tissue removed around the tumor).

Mohs surgery is a layer-by-layer removal of the tumor. Each layer that is removed is examined under the microscope to verify that no tumor remains. If there is remaining tumor in the removed specimen, further excision is done immediately. When the layers are clear, the surgery is finished. This approach allows more of the structure of the ear to be preserved. This technique is particularly useful for recurrent basal cell carcinomas, those larger than 2 cm, or those with an aggressive microscopic appearance. Five-year cure rates using Mohs technique approach 97.1 percent.

Squamous Cell Carcinoma of the External Ear

Squamous cell carcinomas account for 20 percent of all cancerous skin growths (cutaneous malignant neoplasms) and commonly occur in elderly males. Risk factors for squamous cell carcinoma include a suppressed immune system, advanced age, a non-healing ulcer, and a history of exposure to chemicals such as arsenic, soot, coal, tar, paraffin, and petroleum oil. The most important risk factor is exposure to ultraviolet B rays (UVB exposure).

The appearance of squamous cell tumors is variable and includes plaques, nodules, and ulcerations. They may break up into small pieces when rubbed and are prone to bleeding. Squamous cell tumors frequently occur on the helix or tragus, but may occur on any sun-exposed areas.

Imaging may be used to evaluate advanced disease with tumor spread to the adjacent temporal bone and soft tissue structures of the head and neck. The overall risk of metastasis (spread) for cutaneous squamous cell carcinoma of the external ear is approximately 6 to 18 percent, much higher than for basal cell tumors.

Treatment of squamous cell carcinoma of the external ear is similar to that for basal cell carcinoma; however larger surgical margins (the amount of normal tissue removed around the tumor) are required. Some tumors pose a greater risk of spreading disease through the body's system of lymph nodes. Consequently, it is sometimes necessary to remove lymph nodes in the neck (elective neck dissection) and/or to remove the parotid salivary gland for advanced lesions.

The prognosis for squamous cell carcinoma depends not only on the patient's age and overall immune status, but also on the microscopic appearance, size, and location of the tumor. A better prognosis is associated with a well-differentiated tumor (that is, one that is more like a normal structure). The five-year cure rate for squamous cell carcinomas of the external ear range from 75 to 92 percent.

Squamous Cell Carcinoma of the Temporal Bone

The temporal bone is the part of the skull that surrounds the ear. Squamous cell carcinoma of the temporal bone is a rare condition in which the tumor usually spreads into the bone from the skin of the ear canal. The first symptoms may be bloody ear drainage. Some patients may have had chronic ear infections in the past and the diagnosis may be delayed by confusing the appearance of canal tumors with external otitis.

A biopsy of the lesion in the ear canal is necessary to make the diagnosis. Because of the complex three-dimensional anatomy of the temporal bone, its proximity to vital structures, and low survival rates, tumors developing in this region require aggressive management.

Treatment of squamous cell carcinoma of the ear canal is surgical except in those patients not healthy enough to undergo a major operation. Cancer confined to the skin of the ear canal may be treated with excision of the skin of the ear canal with preservation of the eardrum. If cancer invades the bone of the ear canal and/or involves the eardrum, lateral temporal bone surgery is necessary. In this procedure, the eardrum and ear canal are removed. If the cancer has penetrated the eardrum, removal of the middle ear structures is also necessary. Facial nerve weakness or twitching may require removing the involved segment of the nerve and replacing it with a nerve graft from the neck or leg.

Radiation therapy is used as an additional treatment for patients with more advanced lesions or whose disease has spread into the parotid gland or neck. It may also be indicated for inoperable lesions to alleviate patient discomfort.

Squamous cell carcinoma of the ear canal carries a poor prognosis with recent studies suggesting a five-year survival rate of 83 percent for small tumors down to 25 percent for larger tumors. Facial nerve involvement and spread to lymph nodes are likely indications of a poor prognosis.

Melanoma

The incidence of *malignant melanoma* in the United States is 11.1 patients for every 100,000 individuals each year and continues to increase. Ear melanoma accounts for less than 1 percent of all melanomas and has a 10-year survival rate of 70 percent.

Melanomas involving the ear typically present on the helix (rim of the ear) and are colored, usually black. Initially painless, these lesions may change in size, ulcerate, and bleed. A thorough head and neck examination requires attention to lymph nodes in the neck or parotid gland that may occur with regional spread of disease.

The diagnosis of melanoma is dependent on the microscopic evaluation of a biopsy. Evaluation should include a chest X-ray and liver function tests to rule out metastases, since this type of cancer has a tendency to spread through the bloodstream into the liver and lungs. CT scanning and MRI are also helpful in detecting metastatic disease.

The extent of surgical treatment and the prognosis depends on the size, depth of invasion, and microscopic appearance of the tumor. Spread to adjacent lymph nodes or to distant organs such as the lungs and liver predict a poor prognosis.

Removal of surrounding lymph nodes is controversial and may include elective regional lymph node removal and parotidectomy. Recently, sentinel lymph node biopsy has become a well-accepted approach in the management of surrounding lymph nodes with no clinical evidence of spread. One or two lymph nodes in the drainage area of the lesion are biopsied for micro-metastases. If tumor cells are present, the rest of the nodes of the drainage area are removed.

Glandular Tumors

Glandular tumors of the ear canal are rare.

Patients with glandular tumors of the ear canal may complain of ear drainage, ear pressure, pain, and conductive hearing loss. Sensorineural hearing loss may signify tumor extension into the inner ear. Imaging by MRI and/or CT scan is helpful in determining the amount of bony erosion and the size of the tumor.

Benign glandular tumors are treated with excision of the tumor and appropriate margins. Malignant tumors are treated with temporal bone resection (see above for details of surgery), and consideration should also be given to radiation after surgery. Removal of the parotid gland may also be necessary in some cases.

■ A college wrestler suffers a blow to the right ear when taken to the mat by his opponent. He develops tender swelling of the entire ear, except for the earlobe, a few hours later. ■

The external ear (pinna) is subject to a wide variety of injuries, no doubt because it sticks out from the side of the head. In the aftermath of a traumatic event, an ear injury can be highly visible. But all trauma patients must be stabilized before any treatment begins and their injuries must be dealt with based on their severity.

Auricular Hematoma

Auricular hematoma refers to the accumulation of blood in the space between the ear cartilage and its lining (*perichondrium*). It is usually the result of blunt trauma. Cartilage lacks its own blood vessels and instead relies on the blood vessels of the perichondrium to provide it oxygen and nutrition by diffusion. When the shearing forces of blunt trauma cause an accumulation of blood between the cartilage and its lining, the blood supply is cut off. As the cartilage loses its nutrient supply, parts of it die and are replaced with scar tissue. The loss of blood supply can lead to infection and further destruction of the cartilage.

Once the auricular hematoma forms, the ear looks swollen and it may feel doughy or as if there is fluid under the skin. It is often red and shows signs of the initial trauma, like broken blood vessels in the skin. The swelling hides the usual ridges of the pinna. It is essential to drain the blood collection early. Failure to promptly remove the blood clot may lead to infection and/or cartilage necrosis (tissue death) and the permanent disfigurement known as "cauliflower ear."

In the early stages, small hematomas can sometimes be removed with a large-bore needle and syringe. But when the blood is fully clotted or there is a moderate to large accumulation, drainage of hematomas requires a surgical cut parallel with the natural skin folds. Once the skin and perichondrium lining are opened, the hematoma in this space is irrigated copiously with saline spiked with antibiotics to reduce the risk of later infection.

The key to preventing re-accumulation of blood is bandaging the skin firmly against the underlying cartilage over a period of days. Cotton soaked in antibiotic solution is molded to all ridges and valleys of the affected area. The cotton can be held in place by a bulky dressing and head wrap. Other options include sewing cotton bolsters in place with stitches that run through the entire pinna from back to front, silicon putty, and water-resistant thermoplastic splints. Through-and-through sutures without a bolster have also been used.

Lacerations

Sharp or severe blunt trauma may lead to laceration or amputation of the auricle. Immediate repair and prevention of infection are essential. Auricular lacerations should be cleansed of any foreign material prior to repair. The skin edges of simple lacerations can be closed by suturing the edges back together. If some of the tissue has been destroyed, closure may require more complex plastic surgery. This may include advancing or rotating skin flaps from adjacent areas or the use of free skin grafts.

When the entire pinna is cut from the side of the head, it may be placed in a pocket of skin made behind the usual position of the pinna to preserve the cartilage until the acute trauma is resolved and the patient is past the stage of likely infection. The cartilage can then be reattached with microvascular techniques. If the pinna remains hanging by a thread, that thread often contains an artery and, if so, it can be reattached immediately.

Repairs are covered with pressure dressings to prevent swelling and the formation of blood clots, and cartilage-penetrating antibiotics are often prescribed. Excellent cosmetic results can be achieved, even with extensive lacerations.

Frostbite

Freezing temperatures can directly damage skin and cartilage cells as well as block their blood supply. In the early stage, this process may be reversible, but over time, it leads to tissue necrosis (death). Temperatures below 10 °C may lead to numbness of the ear, so people are frequently unaware of impending frostbite. The ear is initially pale and then blue tinged. Ultimately, as the ear thaws, pain, redness, and skin blistering may develop as a result of fluid or blood accumulation.

The initial treatment for auricular frostbite consists of rapid rewarming of the ear using warm, 40 to 42 °C, water. Blisters that are not filled with blood may be opened and patients should be given pain medicine and antibiotics. Even though it may appear that large areas of skin have been destroyed, any removal of apparently dead tissue should be delayed for several weeks until it is absolutely clear which tissue has survived.

Burns

Thermal injury is usually classified by the extent and degree of the burn. Superficial burns involve only the outer layer of the skin (*epidermis*). Partial-thickness burns extend into, but not through, the *dermis* (deeper layer with blood vessels and fat). Full-thickness burns extend through the full thickness of the dermis. Subdermal burns extend into the tissue below the skin including fat, muscle, tendon, cartilage, and bone.

Superficial auricular burns appear red because of blood vessel congestion. These burns are painful. Patients with partial-thickness burns usually have blisters that blanch white on direct pressure and are very painful. Deep partial-thickness burns are associated with less pain, and there may be a layer of dead tissue overlying them (*eschar*). Full-thickness and subdermal burns are often less painful because nerve endings have been destroyed. The wound surface eschar is of varying color, but may be gray or black and charred.

Superficial burns do not scar and may be treated with moisturizing creams. The surface of blisters caused by partial-thickness burns are often removed and antibiotic ointment applied. When not deep, these burns heal without scarring as well. Full-thickness, subdermal, and deep partial-thickness burns of the auricle heal with scarring and contracture (scar deformity) and may be complicated by infection of the cartilage. These burns should be treated with both topical (usually silver-based) and systemic antibiotics (taken orally or intravenously). Early removal of dead tissue and closure with skin grafts may be necessary. Another round of reconstruction is usually performed at approximately one year after injury.

IV | Disorders of the Middle Ear

18 | Perforated Eardrum and Tympanoplasty

■ She said it was meant to be a slap on the cheek but it caught me on the ear. And now I have blood in my ear and trouble hearing—not to mention the dizziness and crazy ringing. After removing the blood and examining me under the microscope, the doctor says there is a hole in the eardrum, a perforation that will probably heal on its own. But if it does not heal, the hole will require surgical repair called tympanoplasty. I won't be making any more jokes about my mother-in-law's new surgically enhanced lips. ■

The *tympanic membrane* (eardrum or TM) is a thin, round structure that is located between the external ear canal and the middle ear (see chapter 1). It plays several important roles. First, it seals off the middle ear from water, debris, and bacteria present in the outer ear canal. In concert with the middle ear bones, the tympanic membrane also amplifies the sound pressure coming into the ear by twenty-two times. When the TM is perforated, middle ear infection and hearing loss are common and all water must be kept out of the ear to prevent infection.

A perforation (or hole) in the eardrum creates a communication between the external, bacteria-filled world and the sterile middle ear cavity. This can lead to recurrent bouts of middle ear infection, especially if the ear is exposed to water, as this will carry bacteria through the perforation into the middle ear space.

Larger perforations of the eardrum can cause substantial hearing loss. There is also a risk of epithelial (skin) growth from the outer surface of the eardrum to its undersurface. This can lead to the formation of a destructive growth called *cholesteatoma* in which skin debris builds up into a cyst that can destroy bony structures of the middle ear.

Causes of Perforation

Infection

The most common cause of perforation is otitis media, infection of the middle ear (see chapter 4). When the infection progresses, inflammation weakens the TM and pressure builds behind it, ultimately becoming severe enough to rupture the drum and allow the fluid to drain. Patients experience this as a severe ear pain that quickly goes away when the drum ruptures and relieves the pressure.

Pressure

Barotrauma (damage caused by changing pressure) is another common cause of TM perforations. Extreme changes in atmospheric pressure—for example, when descending in an airplane or diving underwater—will cause sharp pain, and bleeding from the ear may be noted. Barotrauma can also occur following open-hand blows to the ear where a pressure wave travels down the ear canal and ruptures the drum.

Trauma

People sometimes accidentally rupture their own eardrums with cotton swabs or other objects. Another cause of perforation in children is ear tubes

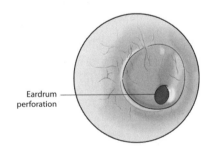

Figure 18.1
The eardrum may rupture because of an infection like otitis media, a penetrating injury (caused by a cotton swab or sharp object), or a compression injury (caused by a slap or other trauma). Most perforations are self-healing, but the larger ones may require an operation called tympanoplasty to repair.

Eardrum perforation

used for chronic or recurrent infection or hearing loss. Approximately 2.5 percent of children who receive tubes will develop perforations.

Self-Healing Perforations

The eardrum has a strong tendency to heal itself. Even eardrums that have been perforated multiple times can still close themselves. It is reasonable to observe a new small- to medium-sized perforation for a period of time to determine if the perforation will close itself. It is critical to maintain strict water precautions during this period of observation, as infection will tend to prevent the drum from healing. There are no exact definitions of an adequate period of observation, but three to six months is commonly used. Larger perforations (greater than 50 percent of the surface area of the drum), perforations with evidence of skin ingrowth, or those which continue to drain infectious material may be considered for earlier surgical treatment.

Tympanoplasty

When holes in the eardrum do not heal on their own it is necessary to perform surgery, called *tympanoplasty*, to close them. By grafting new tissue onto the eardrum the surgeon is able to restore the separation between the external ear canal and the middle ear and recreate a distinct middle ear space. Doing this restores hearing to an optimal status and helps avoid the risk of forming a cholesteatoma (see chapter 20).

There are many techniques used in performing a tympanoplasty, but all of them have in common sealing the perforation with a graft. The graft usually consists of the patient's own living tissue. Once it heals into place, skin of the outer surface of the drum grows over it to complete the repair. Determining the best approach to surgery and the ideal graft material to use is critical to success.

Surgical Techniques

A number of criteria are considered for determining the best approach. The size of the perforation is one critical factor. Tiny, non-healing perforations may be amenable to a simple office procedure in which a small graft is inserted through the perforation or a paper patch is placed over the hole. Total loss of the eardrum may require much more extensive surgery under general anesthesia. The location of the perforation may also be a consideration. For example, certain approaches address perforations in the back half of the eardrum better than perforations in the front half.

The most common technique is the *underlay method* (also called medial graft). In this approach the ear canal skin and eardrum are lifted up and the graft is placed under the perforated eardrum. Grafts are held in place by dissolvable packing. The incision may be limited to the inside of the ear canal if perforations are small and in the back half of the eardrum. In children or adults with small ear canals, a separate incision is necessary in the crease behind the ear. This posterior incision may also be required for larger perforations or those in the front of the eardrum. The underlay technique is fast and effective, frequently taking less than an hour.

The *overlay approach* (also known as the lateral graft approach) is more complex. It involves reconstructing the entire ear canal and eardrum. It is sometimes necessary to use this technique to repair large perforations as well as in revision cases. It typically has the highest success rate in difficult cases, but is technically more challenging and takes longer to perform. Healing time is also longer, typically two to three months as opposed to the one-month period average for the underlay tympanoplasty.

The approach involves an incision behind the ear, but then all of the skin of the ear canal and the eardrum is removed. The bony ear canal is then widened using a drill. A graft is placed over the remaining elements of the eardrum and the skin of the ear canal replaced in such a way that it overlays the edges of the graft. Dissolving packing is placed to hold the graft and skin in place.

Graft Material

A number of different materials, from the patient's own veins to cadaver eardrums, have been used over the years to close eardrum perforations. Fat grafts, cigarette paper, and other man-made substances are still used to repair tiny perforations. But at the present time, the majority of grafts come from the lining of muscle (fascia) behind the ear and are about the size of a postage stamp.

After Surgery

Following surgery, mild pain can be expected. Because the ear is full of packing, a temporary blockage of hearing is likely. After a week or so, patients are usually placed on antibiotic ear drops to dissolve the packing. It is crucial to keep water out of the ear canal when bathing. Swimming is not allowed for six to eight weeks or more. Any remaining packing is removed one month after surgery.

Success Rate

The success rate of tympanoplasty is in the range of 80 to 90 percent. Larger perforations and abnormal findings in the opposite ear are associated with poorer results. Success in adult patients may be slightly better than in children.

Complications of tympanoplasty include:

- bleeding (rare)
- infection (uncommon)
- additional hearing loss (rare)
- facial nerve injury (rare)

Middle Ear Reconstruction for Conductive Hearing Loss

Tympanoplasty for more advanced cases must often be combined with reconstruction of the middle ear bones in order to restore hearing. This is necessary when the bones of hearing do not function properly, a condition that may occur for several reasons outlined below.

Infection

Chronic infection is the most common cause of damage to the middle ear bones. Collapse of the eardrum in the setting of chronic ear infections frequently leads to erosion of the hearing bones, especially at the joint between the incus and the stapes. Infection may also lead to tympanosclerosis or cholesteatoma.

Although commonly seen as a white plaque in the eardrum, tympanosclerosis does not cause hearing loss unless the plaque enlarges and fuses the hearing bones in the middle ear. Patients with this condition usually complain of a progressive hearing loss. Upon examination, the physician will see a white discoloration through the eardrum. Surgical repair has inconsistent results and, if the stapes is involved, there is a risk of sensorineural hearing loss.

Cholesteatoma is discussed in chapter 20. It is a benign, expanding growth that usually occurs in people who have had chronic ear infections. Cholesteatomas are capable of resorbing bone that they come in contact with, often the incus and stapes.

Hearing loss is treated with hearing aids or surgical reconstruction. The surgical procedure consists of microsurgical replacement of the damaged bone. When the malleus or incus has been damaged, various biocompatible prosthetics can be used.

In an outpatient procedure under local or general anesthesia, the connection from the eardrum to the cochlea is restored. The surgery may be combined with tympanoplasty when there is a perforation of the eardrum. When there is no perforation, the surgical approach is similar to tympanoplasty, but without repair of the eardrum.

If the malleus or incus have been damaged, a *partial ossicular replacement prosthesis* (PORP) is used. Usually made of artificial bone (hydroxyapatite) or titanium, these prostheses connect the eardrum to the stapes. When the stapes is also fractured or eroded, a *total ossicular prosthesis* (TORP) is necessary to connect the eardrum directly to the cochlea. A cartilage graft is often used between the prosthesis and eardrum. Success rates for hearing improvement are about 80 percent for PORPs and 70 percent for TORPs.

Traumatic Dislocation

Dislocation of the hearing bones can occur with head trauma or injury of the eardrum with cotton swabs. The most common injury is dislocation of the joint between the incus and the stapes. Repair of this is usually delayed for at least three months after the injury to permit blood and fluid to clear from behind the eardrum. Repair can be accomplished by re-approximating the incus and stapes with a small prosthesis or by performing a stapedectomy if this is not possible.

Congenital Fixation

Fixation of the malleus occurs in about 1 percent of people and causes a mild to moderate conductive hearing loss. Less commonly the incus may be involved. To remedy this the skin of the ear canal and eardrum are lifted as in tympanoplasty and small bridges of bone that fuse the malleus to the surrounding bone are removed. Sometimes a spacer is placed to prevent regrowth in children. This is a highly effective outpatient procedure with a low complication rate.

Congenital anomalies of the stapes represent approximately 40 percent of all congenital ossicular lesions. Although many stapes abnormalities are possible, footplate fixation is the most common. Treatment is identical to that for otosclerosis (see chapter 19).

19 | Otosclerosis and Stapedotomy

■ My hearing started to go about the time Sharon was born and just kept getting worse each year. After two sets of hearing aids (they need to be replaced every four years or so) at $5,000 a pop, I went to see an ENT specialist who gave me a new hearing test and said all of this could be fixed with a simple operation covered by insurance. And yes, I am very unhappy that this possibility was never mentioned previously. ■

Otosclerosis

Otosclerosis is a bone-growth disorder that only occurs in the ear. There are two stages of the process: first the bone becomes spongy, and then it hardens. The most common place for this to occur is where the stapes connects to the cochlea (see figure 19.1). Bone growth fuses the stapes and prevents it from vibrating. Since the stapes won't vibrate normally, sound is blocked and hearing loss results. If otosclerosis penetrates the cochlea, it can also cause sensorineural hearing loss (nerve hearing loss).

Otosclerosis is common. Autopsy studies show that about 3 percent of Americans have microscopic evidence of otosclerosis but that only one in ten of them develop hearing loss. People with hearing loss caused by otosclerosis usually have it in both ears, and it is more common in females than males.

Most people who will develop otosclerotic hearing loss begin to have symptoms between the ages of twenty and thirty, and the loss continues to progress throughout life. It is more common in Caucasians than African Americans, and is rare in Asians. Otosclerosis is autosomal dominant with variable penetrance (that is, there is variable development of disease even if the gene is inherited).

Treatment

Hearing Aids

Otosclerosis is treated with hearing aids or surgery. Oral medications are sometimes used in an effort to prevent the sensorineural component of the hearing loss from getting worse (see below). Hearing aids are a safe and effective treatment, but they have several drawbacks. First off, they are expensive. An average pair of hearing aids might cost about $5,000, and they are generally not covered by insurance. Since they are replaced every four to five years, this can add up to $50,000 or more over an adult lifetime. When you add in $100 per year for batteries you realize what a financial investment hearing aids are. Another drawback is that hearing aids only work when they are in use (for example, not during sleep or while participating in water sports), and unfortunately hearing aids carry a stigma of aging or infirmity.

Medication

Epidemiologic studies have suggested that otosclerosis is less common in areas with fluoride in the drinking water, so fluoride supplements along with calcium and vitamin D have been tried to halt the progression of otosclerosis. Results of fluoride treatment have been at best inconsistent.

Surgery

Stapedectomy is an operation lasting thirty minutes to an hour that is performed through the ear canal under local or general anesthesia, usually as an outpatient procedure. In two variations, the stapes is removed (stapedectomy) or an opening through it is made (stapedotomy).

Lasers are often used to help remove this delicate bone after which it is replaced with a stapes prosthesis. The success rates of stapedectomy and stapedotomy are the same, about 85–90 percent. Efforts to correct

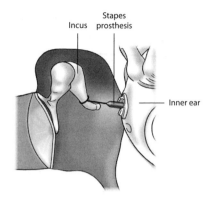

Incus Stapes prosthesis

Inner ear

Figure 19.1
In otosclerosis, the stapes may be immobilized by a bone-like over-growth and must be replaced by a prosthesis to restore hearing.

failed stapedectomy/stapedotomy have lower success and higher compli-cation rates. (Middle ear reconstruction for conductive hearing loss is discussed in chapter 18.)

Risks and Complications

The risk of sensory hearing loss, about 1 percent, is the greatest con-cern to both patient and surgeon in performing stapedectomy. Twenty percent of patients experience a temporary change in taste perception be-cause the taste nerve (chorda tympani) must frequently be moved to per-form the surgery.

20 | Mastoiditis and Cholesteatoma

■ Our son had eleven middle ear infections treated with antibiotics by the time he was two. When we were finally referred to the ear specialist, we discovered that complications had set in—hearing loss, speech and language delay, and worst of all, a growth in the middle ear that is called cholesteatoma. ■

Mastoiditis

Mastoiditis is an infection of the mastoid bone, a hollow, air-containing bone like the nasal sinuses. The mastoid is connected to the middle ear and located just behind the outer and middle ear. Middle ear infections easily spread to the mastoid bone, but usually do not cause clinical problems. The treatment of acute otitis media (see chapter 4) is usually sufficient to also treat mastoiditis if it is present.

However, if acute otitis media does not resolve, it can lead to more advanced clinical mastoiditis. Before antibiotics, mastoiditis was the most common cause of hospitalization of children. About three thousand new cases of mastoiditis are diagnosed each year in the United States, an incidence of about one per year for every 100,000 people.

Symptoms and Signs

Mastoiditis may occur in children and adults who have no history of ear problems. Ear drainage for more than one week is a sign that mastoiditis has developed. It is often accompanied by fever and pain behind the ear in children, but infants may show non-specific signs such as poor feeding, restlessness, and irritability.

Evaluation

Your child's doctor will look in his or her ears and usually find that the eardrum has ruptured and pus is actively draining from the ear canal. The skin of the ear canal may be sagging. The skin behind the ear may be swollen or red and the pinna may be over-protruding from the head. A blood test will show a high white blood cell count, typical of serious infections. Referral to a specialist is quickly made.

The ENT may order a CT scan to determine if the infection has caused destruction of thin partitions of bone in the air-containing mastoid. CT will also identify advanced complications, such as abscess formation and meningitis. If bone resorption (breakdown of the tissue in bones with subsequent release of minerals into the bloodstream) is seen, the term *coalescent mastoiditis* is used and surgery to drain the infection and remove infected bone should be performed within twenty-four hours.

Bacterial Organisms

Mastoiditis is most often caused by *Streptococcus pneumoniae* or *Streptococcus pyogenes*. Other commonly reported causative organisms include *Staphylococcus aureus, Haemophilis influenzae,* coagulase negative *Staphylococcus*, and *Pseudomonas aeruginosa.*

Treatment

Medical and surgical treatment depends on the extent of disease and the condition of the patient. In general, hospitalization is required and intravenous antibiotics are used. If the eardrum is bulging, then the perforation, if present, is not large enough to drain the infection, and a large opening in the eardrum is made. This allows irrigation of the middle ear with antibiotic solution and suction removal of infected material.

If the CT scan shows destruction of the mastoid bony dividing walls, called coalescent mastoiditis, then a mastoidectomy is required. After surgery, a drain is often used to allow the pus to exit through the mastoid incision, avoiding further destruction of the eardrum. Combination suction/irrigation drains, which continuously bathe the mastoid cavity with antibiotic solution, are also used.

Cholesteatoma

Another common complication of chronic otitis media is the development of cholesteatoma. This is a benign cyst-like growth that tends to stay infected. In the process of constantly enlarging, this cyst can destroy bone it comes into contact with.

Cholesteatomas form in chronically infected ears when negative pressure in the middle ear space or ingrowth of the eardrum causes a gradual retraction of the eardrum inwards, forming a pocket. The pocket is lined by skin of the eardrum and, like all skin, it sheds layers every day. The layers gradually build up, enlarging the pocket. This incites a local inflammatory reaction that can erode into neighboring structures, most commonly the hearing bones, leading to a conductive hearing loss. Rarely the cholesteatoma can erode into the inner ear, which can lead to a total hearing loss. It can also cause facial paralysis/weakness and meningitis.

Cholesteatomas are usually infected since the shedding skin serves as an excellent media for bacterial growth. In addition, the bacteria frequently form a biofilm that makes medical eradication of the infection difficult. Bacteria in a cholesteatoma tend to accelerate its growth.

Infrequently, infants and young children are noted to have a white growth behind the eardrum, a cholesteatoma that they were born with.

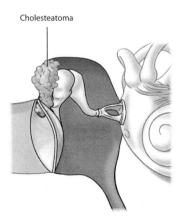

Cholesteatoma

Figure 20.1
Cholesteatoma is a benign growth usually associated with chronic infection. As it enlarges, it destroys adjacent tissue such as the eardrum and bones of hearing.

Congenital cholesteatomas are not related to ear infections. When children with these growths are referred early, the cholesteatomas can be removed simply by operating under the eardrum. However, if there is a delay in diagnosis, the congenital cholesteatoma tends to enlarge and spread and may lead to a lifetime of difficulty.

Symptoms

The most common symptom of cholesteatoma is ear drainage that only temporarily responds to treatment with antibiotic drops. Conductive hearing loss is the other common symptom associated with cholesteatoma. This may be caused by infected fluid or damage to the middle ear bones. Typically, bony erosion involves the incus and stapes while the malleus is less frequently damaged.

Treatment Options

Tympanomastoidectomy

Cholesteatoma must be surgically removed, preferably before bone destruction occurs. Tympanomastoidectomy is an operation that takes two to three hours (or more) to perform. It is usually done on an outpatient basis, but requires general anesthesia. The skin and remainder of the eardrum are lifted to expose the cholesteatoma in the middle ear and an incision is made in the crease behind the ear. Through that incision, the bone overlying the mastoid cavity is removed to expose the cholesteatoma within the cavity. The extent of the mastoidectomy is dependent on the extent of disease and the patient's anatomy.

Intact Canal Wall Mastoidectomy

This procedure leaves the basic structure of the middle ear and mastoid cavity intact. It begins by lifting the remnants of the eardrum and opening the mastoid cavity. The eardrum is reconstructed. There are advantages if the entire cholesteatoma can be removed this way (it usually results in better hearing, and the patient can swim and shower without concern after a healing period). However, because early recurrence of cholesteatoma cannot be detected, a "second-look" procedure is required six to twelve months after the first operation to be certain that there is no recurrence. Erosion of the hearing bones is common with cholesteatoma and frequently requires reconstruction by placement of a prosthesis that reconnects the eardrum with the inner ear (see chapter 18).

Canal Wall Down (Modified Radical) Mastoidectomy

When disease is too extensive or when there is more than one recurrence of cholesteatoma with intact canal wall surgery, the ear canal and mastoid cavity are joined and exteriorized through an enlarged ear canal opening. The eardrum is reconstructed. Modified radical mastoidectomy provides greater surgical access and the ability to see recurrence of cholesteatoma in the mastoid cavity. It does not allow improved visualization of recurrence in the middle ear because the eardrum is reconstructed. Benefits are better visualization and putatively fewer recurrences. Its shortcomings are the requirement for microscopic cleaning of the cavity every six to twelve months throughout life and avoidance of water in the cavity in most instances.

Radical Mastoidectomy

In radical mastoidectomy, as in the canal wall down mastoidectomy technique, the mastoid cavity is exteriorized. However, in this procedure the eardrum is not reconstructed. A very large, sometimes disfiguring, opening from the cavity to the outside is required. This results in worse hearing and often more infections. Radical mastoidectomy is usually reserved for use after failure of more conservative procedures.

Complications

Despite optimal management, cholesteatoma recurrence rates range from 5 to 10 percent in adults and as high as 20 percent in young children. Lifelong surveillance for the disease is mandatory, usually on a yearly basis. As mentioned earlier, patients undergoing a canal wall down procedure will require routine visits to their ear surgeon to clean their mastoid cavity, usually every six to twelve months.

Facial nerve weakness is a rare complication of ear surgery. If facial paralysis or weakness occurs prior to surgery, it is likely that the cholesteatoma has surrounded the facial nerve. Nonetheless, the nerve can most often be preserved. If it is stretched during removal, a temporary weakness may result. If it is cut, the weakness will be permanent and early facial nerve grafting is necessary.

If the cholesteatoma invades the inner ear, removing it may result in nerve deafness and vertigo. Nonetheless, in most cases it must be removed or it will continue to invade, cause deafness, and may lead to meningitis.

V | Disorders of the Inner Ear

21 | Ménière's Disease

■ It hit me like a Bergdorf sack of Jimmy Choos. Spinning, I went down and started to spew, totally mucking my new Armani. My ears were roaring. Mother had buried her third husband with a heart attack, so I knew exactly what was going on. The ER doctor ran a complete heart evaluation, got a CT scan to make sure there was no stroke, drew so much blood that I needed a transfusion, and called in a cardiologist and neurologist. What a shock when all those tests came back normal! ■

This fashion-conscious patient was experiencing her first episode of *Ménière's disease* (MD), but she didn't know it at the time. Ménière's is a sporadic disorder of the inner ear causing attacks of vertigo and loss of hearing. In the full-blown form, it consists of four symptoms, two major and two minor:

- Major
  - Episodes of true vertigo
  - Fluctuating sensorineural hearing loss
- Minor
  - Tinnitus
  - Fullness in the ear.

Ménière's disease has no known cause. However, a related characterization, Ménière's syndrome, refers to the same four symptoms but this term is reserved for use when there *is* a known cause, such as low thyroid levels, Lyme disease, or syphilis.

Ménière's disease is usually first diagnosed when the patient is in middle age, but may be present for many years prior to diagnosis and can begin as early as puberty. The National Institutes of Health estimates that more than 600,000 Americans suffer from MD and that about 45,000 new cases are identified each year. Ménière's disease initially affects only one ear, but many times both ears become involved over a period of decades.

Symptoms

Vertigo

The most disabling symptom of Ménière's disease is vertigo, a feeling of whirling out of control that lasts minutes to hours. Vertigo that lasts only seconds is often due to disorders of the cardiovascular system or benign paroxysmal vertigo (chapter 10). If the spinning lasts more than twenty-four hours, it is more likely stemming from a disorder affecting the brain. Similarly, if there is complete loss of consciousness or there is a shaking seizure, the problem is in the brain or heart. One uncommon variation of MD (called Tumarkin's otologic crisis) is a sudden fall without loss of consciousness.

The spinning vertigo of MD, often accompanied by nausea and vomiting, may occur in attacks lasting minutes to hours. The attacks can occur several days in a row, weekly, monthly, or maybe not for years. Secondary symptoms, the body's response to vertigo, include sleepiness, paleness, and sweating. The first experience of severe vertigo often causes fright or panic leading to an emergency room visit. It is common to conduct a CT scan and EKG to rule out the possibility of a stroke or heart attack.

Sensorineural Hearing Loss

Nerve deafness associated with Ménière's disease is caused by damage to the cochlea and often affects low frequency hearing more than high. This is an unusual pattern. Most ear disorders cause high-frequency hearing loss.

The type of hearing loss is sensorineural (nerve deafness) and it fluctuates or tends to come and go. It is typical for the hearing to get worse for a matter of days, but to mostly improve, leaving a small permanent loss. Over a period of years, these small steps of additional hearing loss can build up and become disabling.

Tinnitus

The tinnitus associated with Ménière's disease also comes and goes and is often compared to roaring or the sound you might hear with a seashell to your ear. Tinnitus may also be a warning, increasing just before an attack of vertigo. It is thought to be caused by a buildup of inner ear fluid pressure.

Fullness

Fullness of the ear can be associated with low-frequency hearing loss or the buildup of fluid pressure in the inner ear. Like tinnitus, ear fullness or pressure may increase before an acute attack of vertigo and provide warning of an attack.

A Typical Ménière's Disease Attack

Many people experience warning symptoms that an attack of vertigo and hearing loss is about to occur. This may be a vague feeling of unease or, as mentioned, a more specific feeling such as tinnitus or fullness. Learning to recognize a coming attack is important because it provides time for the individual to sit, put down the baby, or pull the car to the side of the road.

The intense vertigo of MD is incapacitating. The hearing fluctuates, usually getting worse but sometimes getting better after an attack. The eyes may shake back and forth (nystagmus) due to inner ear balance system nerve connections with the visual system. This causes blurred vision and the vertigo may be reduced by closing your eyes. Many people experience severe anxiety because the symptoms can mimic those of a heart attack or stroke. If you have any doubt it is wise to call 911 for help.

Following the spinning sensation, most people continue to feel nauseated, fuzzy-headed, and sleepy. Common treatments of vertigo (Antivert, Dramamine) also cause drowsiness and most people sleep for several hours after an attack, waking somewhat refreshed.

After many years of episodes the severity of vertigo and fluctuating hearing loss is reduced, leading to what is commonly called "burned out Ménière's." At this stage, constant imbalance may be more prominent than vertigo. Permanent hearing loss is also present in late-stage MD, often requiring hearing aids and sometimes a cochlear implant.

Cause and Mechanism of Ménière's Disease

By definition, MD is *idiopathic* (meaning the root cause is unknown). Speculation includes inflammation (such as would be caused by a viral infection), abnormal circulation, genetics, allergy, and autoimmune disorder. It is also thought that acute inner ear pressure increases can be brought on by caffeine, alcohol, tobacco, salt, and stress (known as CATSS) as well as by anxiety and general illness. As mentioned, in those few patients in whom ear specialists can find a basic cause for MD, the term Ménière's syndrome is usually used.

Regardless of the underlying cause, quite a bit is understood about what happens in the inner ear that results in the symptoms. The basic mechanism is an increase in inner ear fluid called *endolymphatic hydrops*. In some ways, MD is similar to glaucoma, which is a buildup of inner eye fluid. Like MD, glaucoma can cause nerve damage, has an unknown root cause, but can be effectively treated. The two disorders are not related and having one does not dispose you to having the other.

Making the Diagnosis

Medical History

Classical Ménière's disease consists of all four symptoms, but in the early phases only some of the symptoms may be present. Clinically, the diagnosis requires either two major symptoms (vertigo and hearing loss) or one major and two minor symptoms (vertigo or hearing loss plus both tinnitus and fullness). For example, hearing loss and tinnitus may be the earliest symptoms, but the diagnosis of Ménière's disease is not made with certainty unless vertigo or fullness is also present.

Two variations of classical Ménière's disease have also been identified: *cochlear Ménière's* (no vertigo) and *vestibular Ménière's* (no hearing loss or tinnitus). Secondary symptoms include vomiting, hyperacusis (sensitivity to louder sound), poor balance, generalized dizziness, sweating, sleepiness, distortion of sounds, and blurred vision.

When you visit your ear specialist, be prepared to answer questions about your symptoms:

- What is the main reason that you came to the office (e.g., an attack of spinning)?
- What did it feel like? How long did it last? What else did you feel?
- What were you doing just before it happened?
- When was the last time you ate or consumed alcohol before the attack?
- If you have had other episodes, when was the first?
- Describe the duration and frequency of the attacks.
- Do you have any other medical problems?
- Describe your diet, stress level, caffeine intake, and medications.
- Have you been checked by your primary doctor; what tests have been done?
- Does anyone in the family have hearing loss, vertigo, tinnitus, ear fullness, or Ménière's disease?

Even if you fill out a dizziness questionnaire, be sure to tell the doctor or intake nurse your symptoms. The sentence, "Wait, I need to tell you the rest of my symptoms" can sometimes help professionals focus during a busy day. Too often completed forms may be only briefly scanned. Bring notes with you to help you recall details of your symptoms, but avoid long, rambling tales of woe.

Examination

Your ear specialist will check your vital signs (blood pressure, pulse rate, and temperature), examine your ears, nose, throat, head, and neck, and perform a neurological evaluation based on the inner ear. This includes observing your eye movements, standing with feet together and eyes closed, walking in place with your eyes closed, pointing your fingers at the doctor with your eyes closed, touching your fingers alternately from your nose to the doctor's moving fingers, and a variety of other quick and easy checks.

Hearing and Balance Tests

The hearing test (*audiogram*) and basic balance test (*video nystagmogram*, VNG, see chapter 26 for more on balance tests) are nearly always required. The hearing test not only measures your ability to hear, but also the pattern of any hearing loss. Recall that in Ménière's disease, the hearing loss is usually worse in the low frequencies. The VNG measures eye movement caused by warm and cool water (or air) placed in the ear canals. Ménière's disease typically causes a weakness in the response to temperature change, so one ear may have a reduced response.

Other tests can also be very helpful in establishing the diagnosis. *Electrocochleography* (ECoG) measures electrical signals from the cochlea. One particular ECoG pattern is typical of Ménière's disease, but this pattern is present only during active phases, when the inner ear fluid pressure is elevated.

Less commonly, glycerol or urea dehydration tests have been used to determine the potential reversibility of hearing loss in MD. These two drugs are fast-acting diuretics and temporarily reduce the fluid pressure on hair cells in the inner ear. Improvement in hearing confirms the diagnosis of MD and indicates that hair cell damage may not be permanent, that is, hearing may get better with treatment.

Imaging and Other Tests

Imaging studies are often necessary to make sure that other disorders are not the cause of symptoms. CT and MRI are used to rule out stroke, tumors, inflammatory, autoimmune, and other neurological disease. CT is also used to identify another condition that can cause symptoms similar to MD, *superior canal dehiscence*. In this disorder the bone protecting the superior semicircular canal (see chapter 1) is eroded by the pulsations of the brain fluid.

Blood tests including complete blood counts, cholesterol levels, immune system tests, and liver, kidney, and thyroid function studies may be necessary. When the patient's history suggests the possibility of allergy, syphilis, or Lyme disease, these should also be investigated.

Consultation

Your neurotologist may suspect cardiovascular disorders (like abnormal heart rhythm or high blood pressure) or endocrine disease (like low thyroid levels) to be the cause of your symptoms and ask you to consult with your primary doctor, an internist, or other specialist. Primary neurological diseases like multiple sclerosis or stroke would dictate referral to a neurologist.

Treatment

Treatment of Ménière's disease is personalized based on individual needs and the extent of disability. It is focused on reducing inner ear fluid pressure or blocking abnormal balance nerve signals from getting to the brain. Treatment can be categorized into three stages:

Stage 1—Non-invasive Treatment

Stage 1 treatment is aimed at reducing inner ear fluid buildup and is effective in 80 percent of patients. This treatment begins with a low-salt diet (beginning at 1500 mg of salt per day). Find more information about Ménière's diet or low-salt diet by searching these terms on the Internet.

Alcohol should be eliminated because it has its own direct effect on the inner ear as well as the brain. It will also be necessary to gradually eliminate caffeine. If you are drinking more than two cups of coffee per day, reduce your consumption gradually over two weeks or more to limit headache and other withdrawal symptoms. Exercise, proper sleep, and stress reduction are also important in managing this disorder. Your doctor will teach you more about MD—understanding what is going on is very important in controlling it.

Stage 1 treatment also includes the use of oral medications when indicated. The most commonly used medications are mild balance nerve suppressants such as Antivert (meclizine) and Dramamine (dimenhydrinate). These work by reducing the vertigo signals from the inner ear to the brain. Diazepam (Valium) is another vestibular nerve suppressant that also works on the balance centers in the brain, making it more effective for many people than other oral medications. Even though the dose required is lower than that necessary to treat anxiety, diazepam should still be restricted to short periods of use in order to avoid habituation.

Diuretics are sometimes helpful. They are thought to work by reducing fluid pressure. These drugs are safe and complement vestibular nerve suppressant treatment. Patients using diuretics will sometimes need to take potassium supplements.

While diuretics and vestibular nerve suppressants are the mainstay of Stage 1 treatment in the United States, many European neurotologists recommend Serc (betahistine), a vasodilator also used by many American neurotologists. Because its mechanism of reducing hydrops is different than other medications, it can have an additive beneficial effect.

Stage 1 treatment of Ménière's disease is continued for six weeks or longer, unless symptoms are severe. In 80 percent of people, Stage 1 treatment is successful and medications can be tapered off after a period of months. But the restrictions on caffeine, alcohol, tobacco, and stress may need to be continued indefinitely. In the remaining 20 percent of patients, it is necessary to use Stage 2 or 3 treatment methods.

Table 21.1
Summary of Stage 1 Treatment* for Ménière's Disease

- Use a low-salt diet
- Avoid caffeine, alcohol, and tobacco
- Reduce stress
- Exercise (try for thirty minutes of aerobic exercise daily)
- Learn about Ménière's disease
- Take medications
  - diuretic
  - balance nerve suppressant
  - vasodilator
  - oral steroids

*Recommended for most patients.

Stage 2—Minimally Invasive Treatment

Stage 2 treatment is used only if the more conservative Stage 1 treatment does not relieve the vertigo. In most cases of Stage 1 treatment failure, dietary, general health, and medical management is continued while adding Stage 2 options. The availability of minimally invasive methods has revolutionized treatment of severe cases of MD over the past decades by reducing the number of people who require major surgery to control vertigo.

Two of the Stage 2 treatments are administered by intratympanic (IT, that is, in the middle ear) injection. The eardrum is numbed in the ENT office by placing a drop of local anesthetic on it. Next, using a surgical microscope, the doctor uses a tiny needle to inject medication through the eardrum and into the middle ear space.

Intratympanic gentamicin is the most commonly used Stage 2 injection, controlling vertigo in about 80 percent of MD patients who do not respond to Stage 1 treatment alone. However, it is an ototoxic drug that can also cause hearing loss as a side effect. By using small doses and repeating them only if necessary, the rate of hearing loss caused by gentamicin is limited to about 10 percent of recipients. Studies show use of a fixed dose, that is, not a dose tied to the patient's response, causes hearing loss in up to 25 percent of people.

The other commonly used intratympanic medications are steroids. There are several possible mechanisms of action, including reducing inflammation and reducing an immune response. Intratympanic steroids do not cause hearing loss, but are less effective than gentamicin.

Another Stage 2 method is air pressure treatment. A hand-held device pumps pulses of air pressure into the ear canal and through a tube in the eardrum to reach the inner ear. The patient applies the device to the ear for five minutes, three times per day and studies have shown this to be safe and probably effective.

Stage 3—Major Surgery

Surgical management of MD is restricted to patients with frequent, incapacitating attacks of vertigo who did not respond to Stage 1 and

Stage 2 treatment. Going on to Stage 3 treatment is necessary in fewer than 10 percent of MD patients. Once common, these operations are now infrequent.

Endolymphatic Shunt

This is the most conservative surgical procedure, designed to preserve both hearing and balance. It is non-destructive, theoretically allowing excess fluid to drain from the inner ear into the mastoid cavity where it is absorbed into the blood. It is done to control vertigo and is effective in up to 80 percent of patients who find no relief from all Stage 1 and 2 treatments. Hearing improvement has also been reported. This operation takes one to two hours under general anesthesia as an outpatient. Complications are infrequent but include hearing loss and spinal fluid leakage.

Vestibular Neurectomy

This operation consists of cutting the balance nerve between the inner ear and brain. It preserves hearing but destroys the balance function of one ear. The opposite ear takes over during the following weeks. It is a permanent cure for vertigo arising in the operated ear but has no effect on MD that may be present in the opposite ear. Unfortunately, it cannot be performed in both ears because cutting both balance nerves would cause permanent incapacity. This operation has the risks and complications that accompany any brain surgery because it requires access through the lining of the brain and cerebrospinal fluid to reach the balance nerves. It requires two to three days of hospitalization.

Labyrinthectomy

Labyrinthectomy is a one- to two-hour operation to remove the inner ear. This procedure eliminates vertigo coming from the operated ear but destroys both hearing and balance in that ear. It can only be performed on one side and only when hearing is already severely impaired. It is completely destructive but has the highest rate of vertigo control.

In summary, Stage 1 treatment is non-invasive, consisting of low-salt diet, exercise, and stress reduction at first, then adding diuretics, vasodilators,

Table 21.2
Summary of Treatment Stages for Ménière's Disease

Stage 1 (non-invasive)	Stage 2 (semi-invasive)	Stage 3 (surgery)
Diet, reduce stress	IT* Dexamethasone	Shunt
Exercise	IT Gentamycin	Vestibular neurectomy
Oral medication	Air Pressure	Labyrinthectomy

*IT = intratympanic injection.

and balance nerve suppressants when necessary. Stage 2 treatment is semi-invasive, consisting of injections of medication through the eardrum or pressure treatment through a ventilation tube in the eardrum. Stage 3 treatment consists of major surgery and is required in a small minority of cases with disability caused by vertigo.

22 | Noise-Induced Hearing Loss

▨ Samuel didn't think he tuned-in more music than his friends.
But from the time he could walk the boy could dance. If he was
walking, he was bobbing to his player, and you could hear it, too,
when he went by. In class, on the bus, at home he was listening
and moving. And now the doctor tells him he has the hearing of a
fifty-year-old. And it's only going to get worse. ▨

Exposure to loud noise is the second leading cause of hearing loss. *Noise-induced hearing loss* (NIHL) can occur at any age. The National Institutes of Health estimates that approximately twenty-six million Americans between the ages of twenty and sixty-nine have hearing loss that was caused by exposure to noise at work or during leisure activities.

Noise-induced hearing loss in adolescents is growing especially rapidly. It is estimated that 20 percent of US adolescents between the ages of twelve and nineteen have it. Higher rates are found in lower socioeconomic populations. This represents a growth of more than 30 percent over a twelve-year interval.

The louder the noise and the longer the exposure, the more damage it causes to the hair cells of the inner ear (see chapters 1 and 5). For very loud sounds, such as those coming from a jet engine, there is no safe duration of exposure without ear protection. Nonetheless, noise-induced hearing loss usually builds up slowly and painlessly over many years. The exposure is a cumulative but preventable cause of hearing loss. Non–work related activities that result in nerve damage include the use of music players, snowmobiles, outboard motors, lawn mowers, dirt bikes, leaf blowers, and power tools. Listening to loud music at concerts or in a band also plays a large role in permanent hearing loss.

Current music players differ from their predecessors in that they can play a full day of continuous music. The longer duration means that even softer sounds can cause hearing loss. But the problem is not just about the player, it is also the speakers. Earbuds are closer to the eardrum than headphones and can cause more damage. As a general rule, music played through earbuds or headphones should not be audible to anyone else (keep the volume down) and should not be played for more than two hours at a time (keep the duration down as well). Another general rule: if your ears ring after sound exposure, you have already caused some damage.

Occupational Noise Exposure

According to OSHA (the US Occupational Safety and Health Administration), approximately thirty million Americans are exposed to hazardous noise at work each year. OSHA considers noise-induced hearing loss to be one of the most prevalent occupational health concerns in the United States. Tens of thousands of workers suffer from preventable hearing loss each year.

OSHA limits workplace noise exposure to 90 dB for an eight-hour period. For every 5 dB over 90, the duration of exposure is cut in half. For example, if you are exposed to 95 dB, the maximum exposure time is four hours, and for 100 dB, two hours.

As a practical reference, conversation is in the range of 60 dB, heavy traffic 85 dB, music players are often used at over 100 dB. Firearms, which can range up to 150 dB, have no safe usage period without protection. It is an important fact to recognize that the OSHA limits are the result of negotiations between scientists, physicians, audiologists, politicians, and manufacturers. The OSHA exposure limits are purposefully loose enough to allow a small number of susceptible people to develop NIHL. For this reason, it is important to reduce exposure as much as possible rather than to just meet OSHA regulation.

Temporary and Permanent Threshold Shifts

Noise-induced hearing loss can be temporary or permanent. After noise exposure, the hearing may drop considerably but a large portion may return to normal (*temporary threshold shift*, TTS) if the ear is allowed to recover. But not all hearing returns to normal. The hearing that does not return is called a *permanent threshold shift* (PTS). But even in temporary loss, a degree of permanent damage is present and irreversible. Noise-induced hearing loss is cumulative.

How Loud Sound Damages Hearing

Prolonged exposure to noise greater than 85 dB causes damage to hair cells that is visible under electron microscopy (EM). The first finding seen on EM is breaking or fusion of the cilia (hairs) of hair cells. This precedes loss of the hair cell itself and degeneration of the cochlear nerve cell. Alterations of the blood vessels may also occur.

Blast injuries, such as those suffered by members of the military, are considered separately from noise exposure. Although a blast is noisy, it's not the decibel of the sound that creates the damage. Rather blast trauma is caused by a sudden pressure wave that can rupture the eardrum and inner ear membranes, fracture or dislocate the middle ear bones of hearing (ossicles), cause instantaneous disintegration of hair cells, and result in leakage of inner ear fluids into the middle ear.

Noise-induced hearing loss is largely preventable. The first step is to avoid exposure to loud sounds. The second is to use ear protection if you cannot avoid exposure. The third is to move away from the loudspeaker if possible (doubling your distance reduces your noise exposure by 400 percent), and the fourth is to limit the time you are exposed.

How much protection is provided by ear plugs and ear muffs? Plugs and muffs vary in their ability to protect the ear so it can be misleading to generalize. Each device will have an official NRR (noise reduction rate) number in decibels (dB). Many people believe that NRRs are exaggerated by manufacturers and should be treated with some skepticism. Nonetheless, as a general rule of thumb, foam ear plugs provide NRRs of about

Table 22.1
Preventing Noise-Induced Hearing Loss (NIHL)

- Understand how much noise (loudness and duration) it takes to cause NIHL
- Avoid exposure to sound over 85 dB
- Anticipate exposure (Going to a concert? Bring ear plugs in your pocket or purse.)
- Move away from the speakers
- Protect the ears of children and demand safe listening for adolescents
- If you think you are losing hearing, have it tested

20 dB. Quality ear muffs provide NRRs of about 30 dB, and plugs plus muffs provide up to 40 dB. The most common use of plugs plus muffs is on the shooting range.

Built-in Protection

Most of the nerves between the cochlea and the brain carry sound signals to the brain. An exception is the efferent system in which nerves run from the brain back to the ear and help protect the hair cells from noise exposure.

Sound conditioning is a training effect that has been used experimentally. Repeated exposure to low-frequency, non-damaging noise can condition the cochlea to noise trauma. This is also called cochlear toughening, but is an experimental process at the time of writing and is not used clinically.

Medical Treatment

When dealing with recent noise exposure, ear specialists can use a variety of medications to treat hearing loss. The key to success is early diagnosis. Steroids, as well as other anti-inflammatory drugs, injected into the middle ear for passage into the inner ear have proved beneficial.

Antioxidants are another category of useful drugs and include common medications such as N-acetylcysteine, allopurinol, and vitamin C. A relatively new category, anti-apoptotic (apoptosis is programmed cell death as a response to injury) drugs are currently used experimentally to reduce programmed cell death that can be initiated by inflammatory or oxidative processes. In addition, hypothermia and vasodilators have been used.

Loud noise can cause progressive hearing loss without being noticed for months or years. If you don't have ear plugs or ear muffs handy, a large piece of wet tissue or paper napkin stuffed in the ear opening (not down the canal) will give you about 10 dB of protection. (If it is not wet, it will not give you any protection.) Move as far from the noise as practical.

If you experience noise-induced hearing loss suddenly, see an ear specialist and audiologist within a week. While your primary doctor or emergency room physician will prescribe oral steroids until you can see the ear specialist, it may be more effective to receive an injection into the middle ear.

23 | Sudden Deafness and Autoimmune Disease

■ Shakeetha woke up one morning and couldn't hear anything from her left ear. She felt a little better when the doctor took her phone call and told her not to worry, it was probably nothing. A week later there was no improvement so she called again and spoke to the nurse. Shakeetha had a hearing test and her ear specialist diagnosed sudden deafness in the left ear. Treatment resulted in a partial return of hearing. ■

Sudden Deafness

Sudden deafness (SD) is a loss of hearing that occurs over a short period of time, no more than three days. Its cause is often unknown and it affects only one ear in most cases. The audiogram (see chapter 26) shows a loss of at least 30 dB in three consecutive frequencies. Thirty decibels is a mild sensorineural hearing loss, but many times the loss is much worse.

Sudden deafness is considered a medical emergency because it requires immediate attention if the hearing is to be restored. Many people notice sudden deafness when they wake up one morning or try to use the telephone in the deaf ear. Occasionally a pop is heard preceding the hearing loss and symptoms of vertigo and tinnitus are often present. Sudden deafness occurs mainly in middle age and affects an estimated five thousand Americans each year. Fifty percent of people recover some hearing without treatment, but this success rate goes up to 85 percent with early treatment.

> **Table 23.1.**
> **Characteristics of Sudden Deafness**
>
> - Develops quickly (over about three days)
> - Usually occurs in only one ear
> - Is considered a medical emergency
> - Audiogram shows a loss of at least 30 dB in three consecutive frequencies
> - Often responds well to early treatment

Causes

The technical name of sudden deafness is idiopathic (cause unknown) sudden sensorineural hearing loss. Most of the time, as the technical name implies, the cause of sudden deafness is unknown. However, some conditions are known to cause sudden hearing loss, in which case it is not idiopathic. These include:

- Ménière's disease
- Ear infections
- Tumors
- Head injury or diving injury
- Medications
- Exposure to loud noise or a blast
- Neurologic diseases
- Autoimmune inner ear disease
- Sickle cell disease, leukemia

For most people with sudden deafness, no cause is found and viral infection or vascular disorders are suspected. Unfortunately the cochlea cannot be cultured or biopsied for a definitive diagnosis; these standard diagnostic methods have a high chance of making things worse, so the true cause is usually unknown at the time of treatment.

Making the Diagnosis

The history you give the doctor is important, so be thinking about the time you first noticed hearing loss, what you were doing then, and what you had been doing for the preceding several days. Had you gone snorkeling or diving, flown in an airplane, had difficulty clearing the pressure in your ears, heard a pop in your ear, had a head injury, or lifted heavy weights? Have you had a recent viral infection or cold sore? Do you have a history of vascular obstruction such as a prior stroke, loss of vision, atrial fibrillation, or blood clots? Were you dizzy or did you notice ringing in the ear when you first lost hearing? Do you have Ménière's disease or have you experienced a prior hearing loss?

Your doctor will perform a careful examination to be sure you don't have a wax impaction, middle ear infection, neurologic abnormality, or rash. An audiogram will show the extent of the hearing loss and an MRI is often necessary to rule out hearing loss caused by a tumor. This occurs in up to 10 percent of patients with sudden deafness. Blood tests for infectious disorders such as Lyme disease and syphilis as well as autoimmune inner ear disease will also be obtained when indicated.

Treatment

Since sudden deafness may be caused by a number of different disorders, no single treatment works for everyone. If a specific cause can be identified, then treatment is aimed at that cause. Otherwise, steroids are considered the treatment of choice.

Steroids

Steroids such as prednisone and dexamethasone are thought to work by reducing inflammation, edema, and programmed cell death (apoptosis). Because short-term use is indicated, steroid complications are usually not a problem. Be sure to tell your doctor if you have diabetes, chronic infection, peptic ulcers, or a disorder of your immune system.

Once administered only by mouth, steroids are now frequently used orally in combination with injections through the eardrum. These injections are usually nearly painless and increase hearing recovery. Middle ear injection of Decadron (dexamethasone) alone is used in patients with diabetes or others who should avoid oral prednisone.

Hyperbaric Oxygen

Hyperbaric oxygen therapy (HBO) is a method that uses a pressure chamber to increase oxygen levels in the blood. Higher oxygen levels are thought to help damaged tissue recover. In sudden deafness, HBO has been claimed to increase the recovery of hearing when added to steroid therapy. However HBO is not widely accepted as a treatment for sudden deafness.

Antiviral Medication

The herpes family of viruses has been associated with SD and is susceptible to antiviral medication. However, antivirals such as acyclovir may not be effective in treating SD.

Vasodilators

Vasodilators are medications that are intended to increase circulation in order to provide more oxygen and nutrients to injured tissue. However, at this time, studies have not demonstrated conclusively that vasodilators improve recovery from SD.

Autoimmune Inner Ear Disease

Autoimmune inner ear disease (AIED) is a rare disorder resulting in progressive nerve deafness of both ears. It is caused by the immune system mistakenly attacking the tissues of the inner ear. The immune system usually attacks bacteria, viruses, and cancer cells using antibodies and immune cells. But in autoimmune diseases, our immune system attacks one or more of our own healthy tissues by mistake. There are over eighty types of autoimmune disorders, including rheumatoid arthritis, lupus, Type 1 diabetes, and multiple sclerosis. Autoimmune inner ear disease accounts for less than 1 percent of all hearing impairment or dizziness. The causes of mistaken immune attacks are unknown.

Making the Diagnosis

Autoimmune inner ear disease symptoms are a result of inflammation caused by the autoimmune attack. They may be limited to the ear (hearing loss, tinnitus, and vertigo) or may occur in association with other autoimmune disorders such as those mentioned above. When combined with attacks on other organs, ear symptoms may be accompanied by fever, joint pain, eye pain, neurological symptoms, rash, and fatigue. The typical onset is sudden hearing loss in one ear that involves the opposite ear over a period of weeks to months.

The first symptoms of autoimmune inner ear disease may resemble sudden deafness. However, after the initial loss of hearing, it usually progresses and involves the other ear over a few weeks to a few months. The finding of progressive, bilateral nerve deafness is the primary diagnostic criterion for AIED. Other hearing tests may also be performed, but generally do not help make the diagnosis.

Testing

There are also a large number of blood tests available to help in diagnosis. However, none of these are specific to AIED or sensitive enough to direct treatment. Tests may be performed to rule out concomitant autoimmune disorders such as lupus as well as infections such as Lyme disease or syphilis. MRI with contrast is sometimes used to identify inflammation of the inner ear or to rule out bilateral tumors (very rare in this context).

Treatment

The initial treatment of autoimmune inner ear disease is the same as that for sudden deafness: steroid therapy that may be administered orally or by injection through the eardrum. Steroids are effective, so much so that a response in both ears to steroids is considered by some ear specialists to confirm the diagnosis.

Steroid treatment for AIED is usually continued for one month, owing to the high rate of recurrence when it is used for shorter periods. Unfortunately, complications from steroids become more likely the longer they are used. Systemic steroids (oral medications that are spread through the blood) are usually not prescribed for patients with peptic ulcers, diabetes, glaucoma, tuberculosis, and high blood pressure. People with these disorders should be treated with injections through the eardrum because that confines the steroid exposure to the ears.

Even when treatment is continued for one month, hearing loss may recur after steroids are stopped, requiring ongoing use of steroids to prevent nerve deafness. This condition is called *steroid dependency*. In such cases, as soon as the steroids are stopped, the hearing loss returns. After a period of time, steroids may stop working adequately. This is called *steroid resistance*.

When dependency or resistance occur, other medications can be used to slow down or prevent further hearing loss. Methotrexate, a mild anti-immune and anti-cancer agent, has been used to treat steroid-dependent AIED, but with little success. Cytoxan (cyclophosphamide) is a similar, but stronger drug that has also been used with sporadic reports of success. However, it has not been adequately studied and sometimes causes serious complications.

A new group of drugs, biologics or biological therapy agents, are created by fusing parts of two or more genes together. They are designed to block the inflammatory process. Preliminary studies of biologics for treatment of autoimmune inner ear disease show some promise, but require replication and larger studies.

24 | Ototoxic Drugs

The treatment recommended to James for dealing with his lung cancer is a combination of radiation and chemotherapy using the platinum agent cisplatin in combination with a non-platinum chemo agent. Here's the dilemma: James already has a hearing loss and cisplatin may cause further loss. In the end he chooses a better chance for survival with no hesitation. Still, when the first dose of cisplatin causes a large drop in hearing, he asks if there isn't something that can be done. James receives steroid injections through the eardrum, quick and nearly painless, which bring back the hearing to its former levels. Next comes a trial of similar injections just prior to each dose of chemo, to prevent further hearing loss.

Prescription medications can harm the inner ear, causing deafness, tinnitus, and imbalance. The term *ototoxic* comes from *oto* = ear plus *toxin* = poison. Ototoxic drugs are usually used to treat life-threatening cancer or infection and on occasion to treat high blood pressure, kidney failure, or severe arthritis. Kidney failure itself can cause high levels of drugs that are excreted through the kidneys, increasing ototoxicity. Some ototoxic drugs, such as aspirin and quinine in high dosage, cause only temporary hearing loss and tinnitus.

Toxicity occurs when medications damage the hair cells of hearing or balance. These sensory cells are among the most fragile in the body. Up to half of all people receiving cisplatin may lose hearing. And about 3 percent of people taking aminoglycoside antibiotics (gentamicin, streptomycin, tobramycin, kanamycin, and amikacin) will also experience ototoxicity.

Genetics and Ototoxicity

A genetic abnormality that causes high sensitivity to aminoglycoside antibiotics occurs in about one in a thousand newborns. Nearly all people with this genetic defect lose hearing after exposure to aminoglycosides. Research is underway to determine if routine screening for this gene should be performed in people about to take aminoglycosides. The genetic disorder is passed on only by the mother. Unfortunately, in many cases there is no adequate alternative to replace this powerful class of antibiotics.

Symptoms of Ototoxicity

When medications damage the hair cells of the cochlea, the symptoms are hearing loss and tinnitus. Many people become aware of the tinnitus first. Damage to the hair cells of the balance system results in dizziness, imbalance, and vertigo.

Medications That Are Ototoxic

Ototoxic drugs can be categorized by the type of drug and whether the effects are permanent or temporary (table 24.1).

What can be done? Precautions are taken when it becomes necessary to prescribe an ototoxic drug. First, you should be informed about high risks and requested to give your consent to treatment. Second, you should have a hearing test before treatment and have your hearing monitored throughout the treatment. Third, drug levels should be monitored in the case of ototoxic antibiotics to make sure you are receiving the right amount. This is especially true in the case of kidney or liver failure, which may raise drug concentrations to toxic levels.

Monitoring consists of testing the hearing periodically during and after a course of chemotherapy. In addition to the usual audiogram that

Table 24.1
Ototoxic Medications and Their Effects on Hearing

Type	Drug	Effect	Early Symptoms
Antibiotic	Gentamicin	Permanent	Imbalance
	Streptomycin	Permanent	Imbalance, hearing loss
	Neomycin	Permanent	Hearing loss
	Erythromycin	Temporary	Hearing loss
	Vancomycin	Permanent	Hearing loss
Chemotherapy	Cisplatin	Permanent	Tinnitus, hearing loss
	Carboplatin*	Permanent	Tinnitus, hearing loss
Loop diuretics	Furosemide	Permanent	Tinnitus, hearing loss
	Ethacrynic acid†	Permanent	Tinnitus, hearing loss
Anti-inflammatory	Aspirin	Temporary	Tinnitus
	Other NSAIDs‡	Temporary	Tinnitus
Other	Quinidine	Temporary	Tinnitus, hearing loss
	Antimalarials	Permanent	Tinnitus, hearing loss

* Carboplatin is a less ototoxic alternative to cisplatin.
† Ethacrynic acid is highly toxic.
‡ NSAIDs = nonsteroidal anti-inflammatory drugs.

tests the frequencies 250 Hz to 8,000 Hz, audiograms should be done to test the ultra-high frequencies from 8,000 Hz to 12,000 Hz, as these frequencies are the first to show hearing loss. Evidence of hearing loss should trigger an immediate consultation with an ear specialist.

It is thought that increasing hydration and slowing administration of anti-cancer agents results in less inner ear damage. These two modifications in protocol are associated with lower chemo levels in the blood but longer periods of exposure. Hearing may be preserved and slower administration does not seem to reduce success in treating cancer.

Preliminary research has shown that steroids can reduce ototoxicity of some chemotherapy agents. In addition, combination of platinum-agent and non-platinum chemotherapy for lung cancer may allow a lower dose of platinum and reduce ototoxicity. Antioxidants may be effective in reducing hearing and balance loss from aminoglycoside antibiotics.

25 | Acoustic Neuroma and Other Tumors

▨ I had just turned fifty when I started losing my balance. Over the next year it got worse, and then I started to have hearing loss in my right ear. Otherwise I felt great. The kids were settled with families of their own, Oscar and I loved our jobs, and the house was paid for. But at my annual exam the doctor confirmed hearing loss and imbalance and sent me to an ear specialist. The test showed a hearing loss in the right ear, and I couldn't understand words very well. Calling it a red flag because it only affected one ear, he ordered an MRI that showed a tumor. And now he's telling me how this tumor that jumped all over my life is not so bad and why I shouldn't be too worried. ▨

Acoustic Neuroma (Vestibular Schwannoma)

Acoustic neuroma (AN) is a benign tumor of the hearing and balance nerve that grows between the inner ear and the brain. It is technically called *vestibular schwannoma* because it arises from *Schwann cells* (insulating cells) of the balance or hearing nerve. As the tumor slowly grows, at a rate of about 1 mm per year, it presses on these nerves and causes hearing and balance symptoms.

Hearing loss is the most common reason people with acoustic neuroma visit the ear specialist. Tinnitus and imbalance are also frequently reported, but other symptoms can include vertigo, facial weakness, taste disturbance, headache, and confusion. It is estimated that acoustic neuroma affects one to two per hundred thousand people and three thousand new cases are identified each year in the United States.

If not treated, acoustic neuroma may grow large enough to press on the brain and cause serious consequences. On the other hand, some acoustic neuromas do not grow at all. Decisions on treatment are based on the size of the tumor, its growth rate, and the age and health of the patient.

Diagnosis

Most people with acoustic neuromas first notice hearing loss in only one ear. Since the common causes of adult hearing loss, namely aging and noise exposure, affect both ears, one-sided hearing loss is a red flag.

History

Your doctor will seek information about your hearing loss. Come to the visit prepared with answers to the usual questions: when you first noticed the hearing loss; whether it has progressed; whether you also have tinnitus, vertigo, imbalance, pressure in the ear, or numbness of the face. He or she will want to know if you have a family history of tumors or have had damage to the ear caused by infection, surgery, or concussion or exposure to the loud explosive sounds of guns, military ordnance, or even a firecracker.

Examination

Your ear specialist will make sure there are no apparent causes of one-sided hearing loss, such as wax impaction, perforated ear drum, or infection. The doctor will also focus on examining the senses and movement controlled by the cranial nerves, those coming directly from the brain. These include hearing, balance, facial motion, facial sensation, and eye motion.

Testing

The audiogram is usually the first diagnostic test performed. It confirms unilateral (one-sided) hearing loss, the amount and pattern of the loss, as well as the ability to recognize words. Poor word recognition is

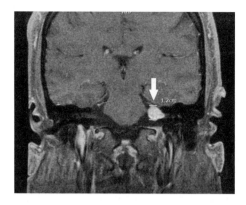

Figure 25.1
This MRI (after intravenous injection of contrast to highlight the tumor) shows a relatively small acoustic neuroma that is barely touching the brain.

associated with acoustic neuroma. A significant hearing difference between your ears requires evaluation with an MRI (see figure 25.1). The MRI has replaced previous batteries of tests and accurately identifies an acoustic neuroma down to a few millimeters in size. When an MRI is not possible, a CT scan with iodine enhancement is substituted.

Management

There are three ways to manage acoustic neuroma: observation, radiation, and microsurgery.

Observation

Acoustic neuromas are benign and usually grow slowly. In elderly patients or those with serious medical issues that would make surgery too risky, observation can be a very good alternative. An MRI is used to accurately measure the size of the tumor and to determine if it is pressing on the brain. It can be repeated in six months, then less frequently if the tumor does not appear to be growing. There is no radiation exposure with an MRI. A sudden drop in hearing or the onset of head pain or imbalance may indicate the need for an MRI at any time. If significant growth of the tumor occurs, the doctor will usually recommend surgical or radiation treatment.

On the other hand, the tumor will never be smaller and the patient never any younger than the first time he or she sees the doctor. If the tumor grows rapidly and treatment becomes necessary, larger tumors and advanced age are both associated with higher complication rates. Patients in their thirties and forties have a long life expectancy and a greater chance for the tumor to begin to grow than do older adults, so a wait-and-see stance (observation) may be less advisable in this age group.

Radiation Therapy

Radiation therapy (RT) can be used to treat small to medium acoustic neuromas. Current forms of radiation therapy (stereotactic radiation therapy such as Cyber Knife and Gamma Knife) are highly focused on the tumor in three dimensions. This allows an effective dose to treat just the tumor while reducing damage to surrounding brain tissue and blood vessels. By delivering small fractions of the total dose over a period of weeks we have a better chance of preserving hearing than if the entire dose is administered all at once. Radiation therapy does not require a hospital stay.

The downside of any form of RT is that it does not kill the tumor but rather slows or stops its growth. RT works mainly by scarring the blood vessels that feed the tumor and obstructing them. Many tumor cells survive radiation therapy and may regrow over a period of years. Another problem is that RT itself is known to be associated with a very small risk (less than 1 percent) of causing cancer. Nonetheless, in a study of acoustic neuroma patients treated with RT, and followed for five years, control of tumor growth was achieved in about 90 percent of patients with AN and 7 percent had serious complications of treatment including hydrocephalus, facial nerve paralysis, and brain necrosis.

Surgery

Surgery for acoustic neuroma is aimed at removing the entire tumor while avoiding damage to the brain and facial nerve. Preserving hearing is also a goal in removing smaller tumors (usually less than 2 cm). Surgery can be used for all sizes of acoustic neuroma, but larger tumors, which are often attached to the brain, facial nerve, hearing and balance nerve, or other nerves from the brain, require different techniques. Removal of larger tumors is associated with more complications. To say that another way, the smaller the tumor is, the better the chance to remove it while preserving hearing and preventing facial nerve weakness as well as avoiding complications.

All of the three surgical approaches commonly used require removing bone and working adjacent to the brain. A stay in the neurosurgical ICU and several days in the hospital are necessary. Full recovery may require several weeks to months and is associated with headache, dizziness, imbalance, and malaise. The three surgical approaches to removal of acoustic neuromas are called middle fossa, translabyrinthine, and retrosigmoid.

The *Middle Fossa* approach is the least invasive but is only used for small tumors. This surgery approaches tumors from above and may allow preservation of hearing. Recovery from middle fossa surgery is often faster than from the other approaches.

Translabyrinthine surgery approaches tumors from the side but always causes complete loss of hearing. It is used for small tumors when the patient has no residual hearing as well as for moderate to large tumors. One advantage to this approach is better visualization and preservation of the facial nerve.

Retrosigmoid procedures approach tumors from behind and are appropriate for all sizes of acoustic neuromas. Hearing may be preserved in patients with small tumors and there is better access to the brain for removal of very large tumors.

A newer technique is being evaluated that uses an endoscope through a small hole in the skull. It has the presumed advantage of being less invasive and early reports show results similar to traditional approaches for small tumors.

Headache is common after surgery for a period of several weeks. In some people, however, it can last for months or years and be debilitating. The retrosigmoid approach is associated with more chronic severe headache than other procedures.

Imbalance and dizziness occur in most patients who have surgical removal of acoustic neuroma. It is usually mild but may be severe enough to prevent their return to work, driving, and other normal activities. Facial nerve paralysis can result from any of these surgical procedures and rarely from radiation therapy. Resulting weakness of one side of the face is a cosmetic calamity. The eyelids won't close and the lower lid droops. The white of the eye turns red with blood vessels showing. The corner of the mouth may droop and drool. Smiling emphasizes the deformity. In mild paresis (partial paralysis), strengthening the facial muscles and re-learning to control them helps many people. In total paralysis, reconstructive surgery and nerve substitution can also ameliorate the disfigurement. In general, surgical complications vary according to the size of the tumor and are more common than complications of stereotactic radiation therapy.

Meningioma

Meningiomas grow from the membranes that surround the brain and spinal cord called meninges. When they occur near the hearing and balance nerve, they may behave like acoustic neuroma. Meningiomas are benign in 90 percent of people. Like acoustic neuroma, meningiomas usually grow slowly and cause problems by pressing on adjacent nerves or the brain. Meningiomas are the most common brain tumor, comprising more than 25 percent. Most meningiomas occur in people between the ages of forty and seventy and are more frequent in women than men. Meningiomas are diagnosed by imaging, either CT scan or MRI.

Treatment

Like treatment for acoustic neuroma, treatment for meningiomas may involve observation, microsurgery, and radiation. Observation with imaging is often recommended for older patients with no symptoms. Up to two-thirds of observed patients have no growth. However, if growth does occur, it is often faster than in acoustic neuroma.

After surgical removal of meningioma, patients experience a recurrence of the tumor at a rate of 10 to 40 percent within ten years. The technique for surgically removing meningiomas affecting hearing and balance is similar to the retrosigmoid approach used for removing acoustic neuromas. Radiation is often recommended as an adjunct to meningioma surgery. In low-grade meningiomas it is used when there is incomplete resection of the tumor. Radiation therapy is also advised in advanced meningiomas due to high recurrence rates.

Glomus Tumor (Paraganglioma)

Glomus tumors (GT) are benign, slow-growing vascular tumors that tend to occur in the middle ear and areas surrounding the ear (see figure 25.2). Unlike acoustic neuroma and meningioma, which affect the nerve of hearing and balance, glomus tumors usually affect the middle ear. Over time, they enlarge by eroding surrounding bone and may cause pulsatile tinnitus, hearing loss, and facial nerve weakness. Rarely, glomus tumors release hormones that can cause rapid heartbeat, flushing, and headache.

Two types of glomus tumors are clinically recognized. When glomus tumors arise from the middle ear they are called *glomus tympanicum*. This name infers small size and lack of involvement of the jugular vein where it runs through the ear. The term *glomus jugulare* indicates that the site of origin is the jugular vein. Glomus jugulare tumors tend to be larger and more difficult to remove surgically due to complex vascular connections and their

Figure 25.2
CT scan showing a small glomus tumor (called glomus tympanicum because it arises in the middle ear).

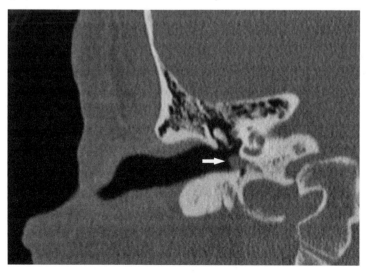

location. Glomus tumors are sometimes inherited and may be more common in people living at high altitudes, although limited evidence exists.

Diagnosis

History

Most people with glomus tumors note pulsating tinnitus that is in synch with their heartbeat. It is caused when the tumor enlarges to touch the eardrum or bones of hearing. Growth of the glomus tumor will also cause a conductive hearing loss when it presses against the eardrum or bones of hearing. Occasionally, people with GT may also experience flushing and palpitations. Before seeing the specialist, check with family members to find out whether anyone else in the family has had an ear tumor.

Examination

A reddish mass is often seen behind the eardrum. Slight air pressure applied to the eardrum may cause the reddish tumor to become pale. In large tumors, major nerves from the brain may be affected so they will be carefully evaluated by your ear specialist.

Testing

Your hearing will be tested and imaging studies ordered. Both CT and MRI scans are useful in determining the size and exact location of the tumor. CT is especially helpful to see if small tumors have surrounded the bones of hearing. If surgery is planned, an *angiogram* (injection of dye into blood vessels) is performed to determine which arteries feed the tumor so they may be tied or closed off (ligated) before removing the tumor. The angiogram also allows *embolization*, a procedure that clots off small capillaries within the tumor.

Treatment

Like other tumors of the ear, glomus tumors may be observed with repeated imaging, removed with microsurgery, or treated with radiation. The best course of action is dictated by the size and location of the tumor as well as by involvement of cranial nerves, the patient's age, and their preference for treatment.

The usual goal of surgery is total removal of the tumor. This is a simple, routine procedure for small tumors but may be impossible in very large tumors that attach to the brain, carotid artery, and cranial nerves. Excision of large glomus tumors often requires multiple blood transfusions.

Radiation is used when only part of the tumor is removed or as the primary method of treatment. The goal of radiation is to stop the growth of the tumor by damaging blood vessels and causing scar formation. Radiation also shrinks the tumor somewhat because it is largely made up of blood vessels.

VI | Other Things You Should Know

26 | Hearing and Balance Tests

When you can measure what you are speaking about, and
express it in numbers, you know something about it; but
when you cannot measure it, when you cannot express it in
numbers, your knowledge is of a meagre and unsatisfactory
kind. —WM. THOMSON, LORD KELVIN, 1883

Hearing Tests

The purpose of audiometry is to measure the amount and type of hearing loss. Audiometric tests can be subjective (based on responses of the patient) or objective (direct measures of function that are beyond the control of the patient). Some tests measure tones, others test for speech recognition. Speech recognition is the larger category, comprised of word recognition and sentence recognition. Tests are often combined to cross check results and improve accuracy.

The two major types of hearing loss are conductive and sensorineural. Conductive hearing loss is caused by a blockage that prevents sound waves from passing through the outer and middle ear. Sensorineural hearing loss, sometimes called nerve deafness, is caused by malfunction of the inner ear (see chapters 1 and 5). Audiometry can help identify which type of hearing loss someone has.

In air conduction testing, sound is delivered through the air using headphones, inserts, or speakers. Sound enters the outer ear, passes through the ear canal, the eardrum, to the middle ear, and finally to the inner ear and brain. Air conduction is how we normally hear, and testing it will show if we have a hearing loss and how bad it is.

Bone conduction, on the other hand, is tested by placing a sound vibrator directly on the skull. The vibrations pass directly through the skull bones to the inner ear. Bone conduction bypasses the outer and middle ear, so if a problem is located there, it affects air conduction but not bone conduction. On the other hand, if bone conduction scores are the same as air conduction scores, then the hearing loss must be coming from the inner ear and the type of hearing loss is sensorineural.

Pure Tone Audiometry

The softest sound that a person can hear (threshold) is measured by presenting progressively louder and softer tones until the loudness that can just barely be heard is established. The goal is to find the threshold at each of six frequencies for each ear. The frequencies usually range from a low of 250 Hertz (Hz or cycles per second) to a high of 8000 Hz in one-octave intervals. Threshold results are displayed on an audiogram (see figure 26.1) and are often summarized by an average of the results at the speech frequencies 500, 1000, and 2000 Hz called the *pure tone average* (PTA). The PTA in figure 26.1 would be calculated at 37 dB.

This patient noticed a loss of hearing in the left ear. The audiogram shows normal bone conduction (bracket symbols) indicating normal nerve function. But it also shows abnormal air conduction (X symbols) and is consistent with a conductive hearing loss such as that caused by fluid behind the eardrum.

This type of test is subjective, meaning that it relies on the patient's own responses to tones. Subjective tests are most common in older children and adults. Another category of test is termed objective, meaning

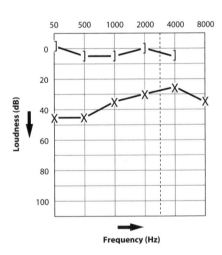

Figure 26.1
Audiogram of left ear shows the softest sound a patient could hear at each frequency. It shows normal bone conduction (*top line*, no nerve deafness) and a conductive hearing loss (*bottom line*, air conduction sound waves blocked).

that responses are directly measured from the inner ear or brain. Objective tests are especially useful in infants and young children who may not be able to respond accurately.

One objective method of determining threshold records brainwaves stimulated by sound. This is called the *auditory brainstem response* (ABR). The sound is presented at different loudness levels to find the softest level that causes a brainwave response. ABR is used to verify subjective findings in adults, for certain diagnostic studies, and for younger children whose developmental level does not allow them reliably to respond when they hear something.

In between infants and older children, a combination of age-appropriate threshold tests is often used that includes objective testing as well as developmentally appropriate behavioral testing. Beginning at around six months of normal development, a child can respond reliably to a sound stimulus and this response is compared to objective results. Such methods include conditioned response, visual reinforcement response, and play audiometry. The audiologist will use a combination of methods to determine your child's hearing level.

Decibels (dB) are units that measure intensity of sound (loudness). The dB measurement is actually a comparison of the patient's threshold with a standard based on normal-hearing people. The ear has the ability to distinguish a huge range of loudness. Because decibels are based on a logarithmic scale they can cover a wide range while keeping the numbers workable. The range of loudness from the softest to loudest sounds measured is typically 120 dB. In terms of sound pressure, the loudest is one trillion (10^{12}) times louder than the softest.

Speech Audiometry

Like pure tone audiometry, speech audiometry measures threshold or loudness, but in addition, it evaluates the ability of the patient to recognize words (speech recognition) and repeat words (speech discrimination). By one year of age, many children can indicate speech recognition by pointing to pictures, body parts, and people. As they develop, children can repeat words from a list. The results are given as a percentage of words correctly

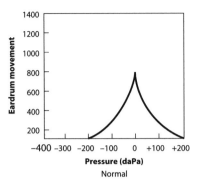
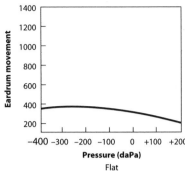

Normal

Flat

Figure 26.2

The tympanogram on the left indicates normal function of the eardrum. The flat tympanogram on the right shows little movement in the eardrum and would be the expected result in a patient with fluid in the middle ear (otitis media).

identified. In older adults, speech understanding may be diminished and this type of hearing loss has been used as an early marker of pre-dementia.

Tympanogram

Tympanometry (immittance audiometry) is a valuable test that can be performed on patients of any age because no conscious responses are required (see figure 26.2). Its volume measurement function can determine whether the ear canal is open or blocked (for example, by ear wax) and indicate the presence of an eardrum perforation. It can measure the stiffness and movement of the eardrum. And it can measure the reflex circuit from the ear to the brain and back.

Otoacoustic Emissions

Otoacoustic emissions (OAE) are sounds that the inner ear produces during the hearing process. These sounds originate in the outer hair cells of the inner ear as they do their job, amplifying incoming sound. Otoacoustic emissions are measured by a microphone in the ear canal and will be absent when the outer hair cells of the cochlea are not working or when otoacoustic emission sound is not transmitted back to the microphone by the middle and outer ear. If the tympanogram is normal (meaning that the ear canal and middle ear are normal) but the person does not have OAEs, that is an indication of an inner ear problem.

Auditory Brainstem Response

The *auditory brainstem response* (ABR) measures the electrical waves of the cochlea and nerves of hearing that pass to the brain. The stimulus is a synchronized series of tones or clicks that in turn synchronize the auditory brainwaves. Electrodes pick up these brainwaves and an averaging computer extracts the synchronized responses of the auditory system from the rest of the brainwaves. Five waves are normally present, the most important being number five. The key results are latencies, the time between the stimulus and the waves. This test can help predict threshold, but can be somewhat unreliable in infants and premature children where it should be confirmed by other tests.

Auditory Processing Disorder Tests

Auditory processing disorder (APD) is an umbrella term that refers to the way the brain processes auditory signals from the cochlea. Even though their ears are normal, people with APD may have difficulties in recognizing and understanding words. Males are said to be affected twice as often as females. The greatest concern is in children with learning disabilities and in older adults with pre-dementia.

Too little is known about the causes of APD, but it can affect the integration of signals from the two ears. This integration normally allows us to locate the source of a sound, combine the information from the two ears, recognize patterns, integrate timing information, understand different words presented simultaneously to each ear, add the loudness from each ear, and pick the clearer signal from whichever ear has less background noise.

Some of the symptoms of APD include difficulty paying attention and remembering oral instructions, trouble with following multistep instructions, low reading skills, and generalized language problems. *Central auditory processing* (CAP) tests can be based on verbal or non-verbal stimuli. APD is associated with ADHD (attention deficit–hyperactivity disorder) and Asperger syndrome and is similar to specific language impairment (SLI) and dyslexia.

Balance Tests

Video Nystagmography

Video nystagmography (VNG) is a test of the balance system that measures eye movements in the dark using infrared cameras. It is used to help determine whether dizziness, vertigo, or imbalance is caused by a disorder of the inner ear. VNG consists of four main tests: response to temperature change in the inner ear, the effect of changing head position, and two tests that measure the ability of the inner ear to control eye movement in following a moving target.

Caloric (temperature) testing provides the most important information. You will be seated in a darkened room and infrared goggles put over your eyes to record eye movements. Each ear will be stimulated with warm and cool air or water, which will make you feel like you are on a merry-go-round. This is a test of the strength of the semicircular balance canals of the inner ear that is measured by the speed of eye movements in response to temperature stimulus. Both ears should be equal in their response. Unequal responses are usually a result of weakness in one ear and will require follow-up by your ENT to determine the cause.

Next most important is the positional test that measures eye movement as a response to change in positioning or final position of the head. Normally, there is no significant eye movement in response to head position. There are a number of causes of positional nystagmus (quick "shaking" back and forth movement of the eyes), but the most common is called *benign positional nystagmus* (BPV). This disorder is thought to be caused by dislocation of tiny crystals in the inner ear and is treated with guided head movements designed to reposition the crystals into a location where they will not cause dizziness.

The two tests of eye movement are designed to identify dizziness or vertigo coming from the brain and are beyond the scope of this chapter. However, it is important to emphasize that vertigo and dizziness can be caused by disorders of the brain and circulatory (heart and blood vessels) systems, although most commonly they originate in the inner ear.

Rotary Chair

Rotary chair testing also assesses the lateral semicircular canal, but it uses slow rotation rather than caloric stimulation. It provides little additional information in most cases, but is useful in identifying disorders where the caloric responses are equal but greatly reduced, where VNG results are uncertain or VNG cannot be performed (younger children and some physically handicapped), to see if the brain is compensating for damage to the balance system, and for early diagnosis of certain types of drug toxicity that can affect the balance system.

Posturography

This test of balance uses a force-measuring platform where a patient stands while wearing a safety harness (to prevent a fall). In a series of challenges to the balance systems (inner ear, visual, and joint position sense), an evaluation of the presence or absence of a balance problem as well as the system causing the problem can be made.

For example, the platform may shift, the visual surround may move (giving a sense of motion like the one experienced at a stop light when a motorist to your side inches forward and you feel like you're inching backward). This information can be helpful in designing rehabilitation programs to prevent falls, especially in the elderly.

Blood Tests

Infection

Syphilis is a sexually transmitted disease that can be congenital (transmitted from mother to newborn) or acquired later in life. In the congenital form, syphilis can cause infantile hearing loss or a hearing loss that shows up only later in life. Adults can sustain hearing loss from the acquired form of syphilis. Sensorineural hearing loss with low speech recognition ability is often associated with imbalance or vertigo (a false sense of movement). When syphilis is suspected, antibody and serological tests are used to confirm the diagnosis. Treatment with antibiotics is highly effective but does not reduce the hearing loss that has already occurred.

Lyme disease is caused by infection with a spiral bacteria (*borrelia*) in the same spirochete family that causes syphilis, but it is not transmitted sexually. *Borrelia* bacteria are present in mice and deer and are transmitted to humans by ticks. It is the most common tick-borne disease in the United States and is prevalent in the northern forests of New England and the Great Lakes regions.

If Lyme disease is not treated, late-stage Lyme disease may develop and cause tinnitus, hyperacusis, or sensorineural hearing loss in up to 40 percent of patients. It may take two or more days for a tick to transmit the disease, so careful washing of skin and clothes after hiking is an effective way to prevent infection. Hot water or a clothes dryer will kill most ticks.

The first sign can be a "bull's eye" rash at the site of the tick bite. This rash expands outwardly. Ear symptoms and facial palsy may not occur for months or even years after infection, during which time a patient may feel symptoms often mistaken for chronic fatigue syndrome and fibromyalgia.

A blood test for Lyme antibodies can be obtained, but oftentimes a trial of antibiotics is used instead. Treatment for Lyme disease has been effective but may be too late to reverse cochlear damage.

Genetic Testing

Genetic testing is part of the routine evaluation of children with hearing loss. Both families and physicians gain important information from these tests that are used to understand the cause and effects of hearing loss and options for treatment and future family planning. Even when knowing the precise cause of hearing loss will not change the course of treatment, families often feel a sense of relief when they understand the reasons it has occurred.

The first gene for non-syndromic hearing loss (no abnormalities other than hearing loss) was only identified in 1997. As noted in chapter 6, a mutation of the *GJB2* gene is known to cause up to 50 percent of non-syndromic deafness. This is the most common cause of deafness in infants.

Routine screening of non-syndromic hearing impaired infants is limited at this time but should include blood tests for the gene *GJB2* along with *GJB6*, a close relative, both of which are causes of connexin-associated deafness. Current research tools allow screening of multiple genes simultaneously and will become widely available.

The finding that a child has *GJB2* deafness provides important information to the parents. They will learn that cochlear implants are highly effective in this population of children; that no other abnormalities (for example, those involving the eyes, kidneys, or heart) are likely to be linked to this gene; and that the chances for future siblings to be hearing impaired are 25 percent.

Specific screening of hearing impaired infants is performed in some syndromic deafness as well. For example, if a CT scan shows malformation of the inner ear (including EVA and Mondini deformity [see chapter 6]), the gene *SLC26A4* is tested for. It can be the cause of inner ear malformations and can also produce thyroid goiter.

Autoimmune Tests

Autoimmune inner ear disease (AIED) is rare, causing less than 1 percent of hearing loss or vertigo and can be isolated to the inner ear or combined with other syndromes such as Lupus, Sjögren's, Behçet's, ulcerative colitis, rheumatoid arthritis, and others. It is thought to be caused by antibodies that mistakenly attack the inner ear resulting in a bilateral progressive sensorineural hearing loss.

The diagnosis is frequently made when a progressive hearing loss responds to steroid treatment. Although many general immune tests have been used, none has been specific enough to base AIED treatment/no treatment on.

Imaging

CT Scan

Computed tomography (CT, sometimes called "CAT") scanners use X-rays to make highly accurate cross sectional scans of the body's inner structures. The scanner itself is donut shaped or has a rotating arm that moves with each section of the scan creating a series of thin, consecutive cross sections (tomography). In evaluating disorders of the ear, CT is used primarily for bony detail and MRI for soft tissue detail, although there is much overlap. For example, a CT scan is very useful in seeing the bony structure of the cochlea, but not as useful as MRI in seeing the nerves, brain, and fluid spaces in the skull.

Contrast agents (a substance used to enhance the contrast between structures of fluids) are often given to improve the quality of the images and to identify those tissues that tend to absorb contrast. These agents can cause a problem for people with allergy to iodine or whose kidney function is not normal. The more common reactions (occurring in about 1 percent of people) are mild nausea and an itching rash.

Because CT uses X-rays (radiation) to image internal structures, there is a small risk of cancer, especially when CT is repeated on multiple occasions. It is estimated that one CT scan may expose a person to the same radiation as a hundred conventional X-rays.

MRI

Magnetic resonance imaging (MRI) is an imaging technique used by radiologists to show both the structure and the function of body tissues. It can provide an accurate image of tissues without exposing the body to radiation. MRI is especially useful in identifying abnormalities in soft tissues such as nerves and brain as well as displacement of spinal fluid by nerves.

In helping to diagnose the cause of balance or hearing loss, MRI is especially useful in identifying tumors such as acoustic neuroma, meningioma, glomus tympanicum, and facial nerve tumors (see chapter 25). In cochlear implantation, MRI is valuable in identifying the presence of the hearing nerve and obstruction of the fluid spaces of the cochlea. Intravenous injection of contrast (such as gadolinium) is often used to enhance specific tissues and help identify abnormalities.

The MRI magnet aligns the protons (parts of an atom) of hydrogen molecules in tissues and radio waves are used to activate (spin or energize) the protons. Different tissues are activated to different degrees and create a signal that is detected by the MRI receiver. The signal is processed as a digital series of cross sections. Always ask for a copy of your MRI or CT on a CD (compact disk) before leaving the radiologist's office and hand carry it to your ENT to avoid communication failures and a wasted visit.

The process of MRI scanning may cause claustrophobia in some individuals since the standard MRI machine requires that the patient lie in a sort of tunnel. This anxiety can sometimes be relieved with mild sedation. Open MRIs, which alleviate some of the sense of being enclosed, are available in some places, but they are not as powerful as standard MRIs. People with implanted metal objects may not be able to undergo MRI. Examples include cochlear implants (exceptions are possible in specific cases), pacemakers, metal clips from surgery, shrapnel, insulin pumps, prosthetics, etc. In addition, metallic dyes from tattoos, including permanent makeup, can cause irritation to the skin or eye during an MRI. People with kidney disease may not be able to receive gadolinium contrast.

27 | The Ear and Scuba Diving

▨ Jane flew to a resort in Mexico for a vacation. She chose the resort partly because it offered scuba diving lessons and she had been wanting to learn to dive. After a couple of days of basic instruction in a pool, she and the other beginners were taken a little way offshore for an open water training dive. Jane had trouble clearing her ears on descent and was falling behind the class, so one of the instructors helped her catch up by pulling her deeper under water. Jane fought her way to the surface and back to the boat, where she told the captain she was having trouble hearing and had severe pain in both ears. Back at the resort she was treated with nasal decongestant and steroid sprays along with oxycodone for pain. She recovered and was able to fly home at the end of the week, vowing to try rafting next year. ▨

After the first amphibians crawled up onto a beach, the ears of land-based animals had to evolve to function in air rather than water. Fortunately for people who love the water, the human ear has retained the ability to do both. But the ear is a delicate organ and, for many of us, is not efficient at adapting to rapid pressure changes that occur in scuba or free diving. In fact, the most common medical problems of divers involve the ears and occur because the pressure on our ears increases as we go deeper.

Although no one know for sure, it is estimated that there are about four million registered scuba divers with an estimated two hundred thousand new divers trained annually. With that many people in the water, medical complications of the ear may be expected to occur.

Equalizing Pressure in the Middle Ear

Have another look at figure 1.1 in chapter 1. Notice how the eardrum is an airtight seal between the outer world and the middle ear. As pressure in the outer world increases, it will painfully stretch the eardrum inward or even rupture it unless the pressure inside the middle ear is increased to match the outside pressure.

Chapter 1 describes the Eustachian tube, how it is closed at rest but opens when we swallow, yawn, or increase nasal pressure. This normally occurs many times each day and may be accompanied by a clicking or popping sound. The problem with diving is that the pressure change is greater and occurs faster than on land. Surprisingly, equalizing pressure is harder from 0 to 10 feet than it is from 40 to 50 feet. That is why it is so important not to let the pressure get ahead of you. Equalize before you feel discomfort and continue to do so every few feet all the way down.

Methods

Swallow

As noted above, the most natural method of opening the Eustachian tube is to swallow. This is easier on land, so that is where you should begin your practice. Swallow a sip of water and listen for a click or crackling sound. If you're not sure, pinch your nose when you swallow the water. Hint: if you are going to take scuba lessons or learn to free dive, practice equalizing your ears on land, in front of a mirror, until you can always click open your Eustachian tubes. Try it in the pool with your head barely under water.

Table 27.1
The Effects of Pressure on the Ear at Shallow Depths

Depth	Effect
1 foot	Slight feeling of ear pressure
4 feet	Significant eardrum stretch, pain, difficulty equalizing
8 feet	Tissue damage, Eustachian tubes lock—inability to equalize
10 feet	Severe pain, likely ruptured eardrum, vertigo, permanent hearing loss

Pseudo-Valsalva Maneuver

The pseudo-Valsalva maneuver consists of pinching the nostrils and blowing air pressure against them through your nose. Many unsuccessful divers think they are increasing nasal pressure when they are actually increasing pressure in the lungs by bearing down. You fall into this category if your face turns red when you try it. The important thing is to get the pressure to the right place.

The best way to accomplish this is to "make balloons." While learning this technique, look in a mirror, pinch your nostrils at their lowest edge, and blow gently until you can see the nostrils balloon out. Try it again with your mouth open. When you see your nostrils balloon out, the pressure is in the right place and you can add more pressure until your Eustachian tubes gently click open.

Toynbee Maneuver

This technique consists of pinching the nostrils and swallowing (see above).

Lowry Maneuver

The Lowry maneuver combines the Valsalva and the Toynbee. With the nostrils pinched and a little water in your mouth, make balloons and swallow at the same time. The Lowry maneuver is the most difficult to learn, but can be the most effective for many people.

If you are still unable to clear your ears, it might be best to find another sport. Or, you can see an ear specialist. Steroid and twelve-hour decongestant sprays are safe if used as prescribed and sometimes a severely deviated septum can be corrected to improve nasal air-flow and equalization.

Barotrauma

Most ear disorders associated with diving are caused by pressure. Called *barotrauma* and sometimes "squeeze," pressure can affect the outer, middle, or inner ear. Fortunately, most of these problems can be avoided but rarely, complications may be severe or life threatening.

Middle Ear Barotrauma

Middle ear barotrauma is the single most common medical condition of divers. As described above, it occurs when the pressure outside the eardrum is greater than the pressure inside. It is avoided by learning to equalize pressure, not diving when you have a cold or allergy symptoms (sneezing, runny nose), and discontinuing the dive if you are not able to equalize at a depth of 4 feet. If you feel pain at any time, stop and ascend just a few feet and clear your ears. Pain is your warning, do not proceed. If you are having trouble, descend on a dive line or anchor line, hand over hand to control your rate of descent. Go feet first.

The complications of middle ear barotrauma include bleeding into the eardrum or the middle ear, filling of the middle ear space with mucous that can last for weeks, and perforation of the eardrum. If your drum perforates, water will run in through your ear and down your throat as well as cause severe vertigo. You may not be able to tell up from down. Your eyes will be closed because of the vertigo and you may end up swimming downward. Practice taking a deep breath and remaining motionless to allow your buoyancy to carry you slowly to the surface.

Your physician will advise you after an examination, but mild middle ear barotrauma is generally treated by not diving for a period of up to a few months. Most perforations will close without surgery, but the ear must be kept meticulously dry to avoid infection. If the eardrum does not heal in three months, surgical repair may be needed (see chapter 17). Another complication of middle ear barotrauma is inner ear barotrauma.

Inner Ear Barotrauma

Inner ear barotrauma is usually caused by middle ear barotrauma. Pressure on the eardrum is transmitted to the membranes and fluids of the inner ear and may cause membrane rupture. Most membrane ruptures inside the cochlea resolve with bed rest and oral steroids. However one type, perilymph fistula, may require other treatment.

Perilymph Fistula

A sudden pressure wave transmitted to the cochlear fluids may result in rupture at the round or oval window (see chapter 1). This may cause leakage of inner ear fluid (fistula) into the middle ear resulting in nerve deafness, tinnitus, and severe vertigo. It is treated with bed rest and often resolves spontaneously. However, in cases of severe to profound deafness or persistent vertigo, early surgery is recommended to repair the membrane. If necessary, this thirty-minute operation is done through the ear canal under local or general anesthesia as an outpatient.

Decompression Sickness

Decompression sickness (DCS) can also affect the inner ear. Also called the bends, DCS results from ascending to the surface too quickly. Nitrogen, which has been absorbed in tissue due to the high pressures at depth, will go back into the bloodstream and form bubbles when the pressure is reduced too quickly. The nitrogen bubbles can obstruct tiny capillaries anywhere in the body, especially the joints and nervous system. As opposed to barotrauma, which occurs during descent, DCS occurs during or after ascent with no history of difficulty clearing the ears. In sport diving, the depth and bottom time are regulated to prevent DCS. Nonetheless, a five-minute stop at 15 feet is recommended due to the remote possibility of the bends. Working divers, who go deeper and stay longer, may require decompression stops during ascent to allow the bubbles to dissolve.

28 | Airplane Ear

⬛ Baltimore to Chicago to see the grandparents was usually a nice flight. There had been no problems at all on previous trips—until this landing. Our four-year-old woke up screaming and we could feel the eyes of the other passengers from three rows back. We could tell how much pain Annie was in, but what could we do? ⬛

Airplane ear (middle ear barotitis) is painful and also causes temporary hearing loss and ear fullness. It may affect one or both ears. Airplane ear results from air pressure changes while flying, usually during descent and landing. It generally clears on its own but may persist for several minutes to several days after a flight, and occasionally it can result in more serious problems.

Causes of Airplane Ear

During takeoff and the climb to cruising altitude, air pressure in the cabin is gradually reduced. As the plane rises to cruising altitude of about 35,000 to 40,000 feet (about seven miles) the cabin pressure is kept at the equivalent of 8,000 feet. Pressurization increases comfort and safety but is still much lower air pressure than most takeoffs or landings.

During landing, cabin air pressure must gradually increase again. And as cabin pressure increases, the air pressure of the middle ear must increase equally or the eardrum is forced inward.

For most people, pressure is automatically equalized because the Eustachian tube opens during swallowing, chewing, or even talking. But when the Eustachian tube does not open, pressure is not equalized and airplane ear occurs. Airplane ear usually happens during landing and is rare during takeoff.

Adults and children are more susceptible to airplane ear if they have cold or allergy symptoms that cause swelling of the nasal and Eustachian tube membranes. This narrows the Eustachian tube opening and makes it harder to equalize pressure. Some people are born with small Eustachian tubes and others may have chronic ear infections or enlarged adenoids that are also associated with swelling of membranes.

Preventing Airplane Ear

Airplane ear resolves on its own for most people within minutes to hours after landing. Nonetheless, it can be painful and preventative steps can be taken. If possible, avoid flying with a cold, allergy flare-up, runny nose, sneezing, and other upper respiratory problems. If you must fly, talk to your doctor in advance. He or she may prescribe steroid nasal sprays for five days or oral steroids for a period of two days or so. Your doctor may also recommend the use of long-acting decongestant nasal sprays the day of the flight. These nasal sprays improve Eustachian tube function of flyers and are best taken thirty to sixty minutes before landing. Long-acting oral decongestants containing pseudoephedrine should be taken before takeoff and repeated two hours before landing. Be sure to check with your doctor before taking any medications and especially before giving decongestants to young children.

Strategies for Adults

Practice clearing your ears before air travel. Swallowing a sip of water, chewing gum, and yawning all cause the Eustachian tube to open and you may be able to hear a popping or crackling sound. For most people, this step is enough. You can also practice the methods described in chapter 27 that gently force air through the Eustachian tubes and into the middle ear. These are more effective techniques and are used by scuba divers.

Table 28.1
Preventing Airplane Ear in Adults

- Avoid flying with a cold or active allergy if possible
- Learn to equalize your ears before travel. Practice clearing your Eustachian tubes.
- Take oral decongestant medications before takeoff and two hours before landing
- Spray your nose with a long-acting decongestant at least thirty minutes before landing
- Begin to equalize at least thirty minutes before landing (the fasten seatbelt sign will trigger you)
- Equalize your ears before you feel pressure and continue every few minutes until landing
- Try pressure gradient ear plugs

Get ready for descent and landing by beginning to clear your ears at least thirty minutes before landing (about the time you return your tray tables and seatbacks to their original locked and upright positions). Adults should clear their ears every minute or so during descent because it is easier to keep the Eustachian tubes open than to open them under pressure. Eustachian tubes tend to "lock shut" when there is too much pressure, so begin equalizing *before* you feel ear pressure.

Some flyers use pressure gradient ear plugs (such as Ear Planes), which have micro filters that purportedly slow down the rate of pressure change reaching the eardrum. They do not entirely block air pressure from entering the ear canal.

Strategies for Children

Some of the methods above apply to children as well. Be sure to check with your child's doctor before using any medications. Certain nasal sprays and oral decongestants may not be appropriate depending on age and other health concerns.

It is usually best to awaken a sleeping child before descent and to plan infant feedings so that your child is hungry just before descent. Suckling or bottle feeding infants in a semi-upright position is an excellent way to open the Eustachian tubes during descent. Snacking, sucking on hard candy, drinking, or chewing gum can be useful in older children. (Drinking adult beverages may also be advised for parents if these techniques don't work for their children.)

Based on prior experience some parents give their child a pain-relieving medication, such as acetaminophen, two hours before landing, just in case. Be sure to check first with the doctor. If your child has a history of ear problems when flying, your doctor may prescribe ear drops that contain a topical pain reliever. If recommended, they are most effective when warmed slightly. One way to warm drops on a plane is to place the bottle of drops in a cup of hot water for a few minutes. As with baby's milk, you can test the temperature on your wrist, where it should feel slightly warm.

Warmth has been known to relieve ear pain since the time of the ancient Greeks (who used warm olive oil). Ask the flight attendant for two small, moist, warm towels (like the kind they hand out in first class before meals) and two styrofoam cups. (Warm wet paper towels can be used as well.) Place one towel in each cup, spill out any water, and test the temperature to be certain that it will not scald the child. Firmly hold the cups up to the skin around the pinna after you are certain that the towels are not too hot. Warm towels may also be held directly against the pinna if no cups are available. A commercial product called Ear Ease is commonly available online. Simply fill it with warm water according to directions and hold it against the ears.

Complications

Most adults and children who experience airplane ear are symptom free by the time they arrive home from the airport. However, if pain, stuffiness, or hearing loss persists, it is best to see your doctor. Mild complications include a fluid buildup behind the eardrum. This fluid is usually clear, but may contain blood if the eardrum was severely stretched. If the Eustachian tube is blocked, it cannot drain mucous or blood from the middle ear and a temporary hearing loss occurs. Middle ear fluid clears on its own in most cases, but in rare situations may need to be removed.

Perforation of the eardrum, permanent hearing loss, perilymph fistula, tinnitus, and vertigo are also possible complications of airplane ear, but are rarely experienced. For more information, see chapter 27 on scuba diving.

29 | Bell's Palsy and the Facial Nerve

Charlie had heard of Bell's palsy before, even knew someone who had it once. So when the left side of his face started to droop he didn't panic. Over the next two days he couldn't close his eye or keep the left side of his mouth from drooling, but otherwise he felt pretty good. Still, he thought he should see the doctor.

The facial nerve is one of the twelve cranial nerves that arise directly from the brain. It runs in a long bony canal (Fallopian canal) through the bone surrounding the ear to innervate the muscles of the face. Disorders of the ear or the facial nerve itself may result in facial weakness or paralysis.

Bell's palsy takes its name from Sir Charles Bell, who described the symptoms in the eighteenth century. It is characterized by partial or total weakness of the facial muscles. The cause is unknown. The overall incidence of Bell's palsy is estimated at twenty to forty people per hundred thousand and about sixty thousand people are affected each year in the United States. The rate increases significantly when pregnancy, diabetes, and family history of facial paralysis is a factor.

Most people recover completely from Bell's palsy, especially if the weakness is partial. But over 15 percent end up with some degree of permanent paralysis. Usually only one side of the face is involved, but paralysis may affect both sides and may recur on either side.

In addition to facial weakness of unknown cause (Bell's palsy), there are several known causes of facial weakness. These include stroke, pene-

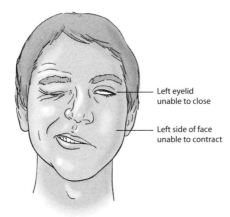

Left eyelid
unable to close

Left side of face
unable to contract

Figure 29.1

Bell's palsy causes mild to severe disfigurement, but 85 percent of patients with Bell's recover fully without treatment. The patient in this illustration is attempting to contract the muscles on the left side of his face.

trating trauma that may damage the nerve, and cancer. But the cause of facial paralysis is known in only about 5 percent of all cases and none of these is Bell's palsy.

What Causes Bell's Palsy?

The facial nerve has the longest course in a bony canal of all the nerves in the body. At one particular place near the inner ear, the facial canal is very narrow and constricting. It is thought that when a virus infects the facial nerve, as it can with any nerve, swelling in this narrow part of the canal causes enough pressure to cut off circulation and result in paralysis. Some people with Bell's palsy have simultaneous infection of other cranial nerves, but those nerves do not stop working because they have room to swell. The viruses associated with Bell's palsy include Epstein-Barr and Varicella-Zoster, both members of the herpes family. Bell's palsy is not related to stroke.

Symptoms

The symptoms of Bell's palsy peak within seventy-two hours of onset. Because the facial nerve has many branches, the symptoms vary widely. For example, besides innervating the muscles of the face, the facial nerve is a path-

Table 29.1
Symptoms of Bell's Palsy

- Facial weakness or paralysis
- Twitching of facial muscles
- Drooping eyelid and corner of the mouth
- Dryness of eye or mouth
- Drooling
- Tearing from the eye
- Metallic taste from the front and side of the tongue
- Pain near the ear
- Tinnitus
- Dizziness

way for tearing, saliva, taste, and for contracting a small muscle in the middle ear. Symptoms of dysfunction may be seen in any of these areas.

Diagnosis

Since Bell's palsy is a diagnosis of exclusion, possible known causes must be ruled out before determining that the paralysis has no known cause. Unfortunately, because 85 percent of patients with Bell's recover fully without treatment, investigation into treatable causes is frequently put off, resulting in a delay of treatment in those cases where it is needed. Other conditions that cause facial paralysis include Lyme disease, sarcoidosis, shingles, acoustic neuroma, facial nerve or brain tumors, ear infections, mastoiditis, cholesteatoma (a growth of the middle ear), and skull fracture.

The history and examination will focus on the onset of symptoms and signs of underlying causes. Special attention should be paid to eye closure and the cornea. When indicated, a CT scan, an MRI, and blood tests can be performed to rule out most underlying causes. Nerve conduction testing can provide information about the likelihood of not recovering from paralysis.

Treatment

The most important immediate need is to protect the cornea. When the eyelids do not close due to paralysis of the face, foreign material may have easy access to the cornea and lead to trauma or infection. Even a pillowcase can pose a danger during sleep with the eyelids open. The cornea quickly dries out when we cannot blink to lubricate it, causing itching, pain, or blurred vision. This may lead to permanent damage.

Moisture chambers and taping or patching the eye are necessary in advanced cases, along with ointments at night and frequent use of artificial tears during the day. The eyelids may need to be taped shut with special tape at night. An ophthalmologist can be very helpful in protecting the cornea and may need to sew the eyelids partially or totally shut if the cornea has been damaged.

With no treatment, 85 percent of people with Bell's palsy begin to have some improvement within three weeks, but only 71 percent eventually regain normal facial movement. This is not deemed a good outcome, considering the nature of the deformity.

Steroids are effective in treating Bell's palsy. Prednisone is commonly prescribed to be taken orally in tapering dosages over ten days. The initial adult dose is usually 40 to 60 mg per day; for children, 0.5 to 1.0 mg per kilogram per day. For maximum effect, treatment should begin within seventy-two hours of the onset of symptoms.

Antiviral medications, although effective in treating herpes-family viruses, have led to only small improvements in patients with Bell's palsy. This may be due to delay in beginning treatment.

Surgical treatment of Bell's palsy is only considered for people who experience complete paralysis and when nerve conduction tests indicate a poor prognosis. Early surgery is aimed at removing the bone from around the nerve to allow it room to swell. The conundrum is that this surgery, to be most effective, should be performed within two weeks of onset of symptoms. But it is not possible to be certain that surgery is necessary during that time, since this is the period when spontaneous recovery also occurs. The most effective decompression surgery is performed through the middle cranial fossa and usually requires a stay in the neuro-surgical ICU as well as several days in the hospital.

If the paralysis becomes permanent, two surgical options may be possible. These include substituting another nerve for the facial nerve or performing cosmetic procedures that are similar to a face lift but with movement added. Gold weights may also be placed in the upper eyelid so gravity can close the eye at rest.

Synkinesis is a complication in which separate branches of the facial nerve fire simultaneously. For example, the eye may blink when you try to smile, giving the appearance of a grimace. The facial muscles are prone to spasm, and crocodile tear syndrome may occur in which tears form when you are eating. The latter symptoms are caused by regenerating nerve fibers following the wrong pathway.

Facial exercises are not effective when the face is completely paralyzed. However in partial weakness or when the muscles first begin to recover, facial exercises can strengthen the muscles and may help separate the function of the eye from the rest of the face.

30 | **The Future** | Gene Therapy and Stem Cell Therapy of the Ear

▓ It's 1970, and everyone knows that there is no treatment for nerve deafness. But by 1984, the US Food and Drug Administration approved the first cochlear implant, leading to advanced computerized devices that let deaf people hear and understand speech. Thirty years later there are clinical trials to see whether gene therapy can restore hearing without implants or can improve implant function. And stem cell implantation may be just a few years behind. ▓

Gene Therapy

Genes are the basic units of inheritance, passed from parent to child. They are located on chromosomes and contain codes for producing specific proteins. These specific proteins determine the structure and function of the body, basically making us what we are.

Gene therapy uses genes that are modified by scientists to prevent or treat many types of disorders, including hearing loss. If a gene has stopped working properly, in theory it can be replaced with a working gene. Alternatively a new gene can be manufactured that will create improvement of function. Ongoing trials examine restoration of hearing by insertion of new genes into the inner ear. The goal is to create new hair cells or to improve function of existing hair cells.

Gene therapy for deafness is investigational at the time of writing, but research has advanced to the point of human research trials. Although promising for the treatment of genetic disorders, some types of cancer, blindness, and deafness, it has not yet been proven safe and effective.

To move new genes into cells, a transporting vector is used, most commonly a safe virus that cannot reproduce itself. Scientists attach genes to viral vectors that are known to be safe and have the ability to penetrate into cells. The vector enters a cell and deposits the new gene. The new gene causes the cell to make functioning proteins to replace the faulty proteins that cause disease. Clinical trials have shown promise in the treatment of leukemia, hemophilia, and retinitis pigmentosa.

Animal studies show that gene therapy can regenerate auditory nerve cells. While the goal of therapy is to regenerate hair cells and promote natural hearing, a preliminary goal of auditory nerve cell (not hair cell) regeneration may be easier to accomplish and improve function with cochlear implants.

The first human trial of gene therapy for deafness is now underway. It treats deaf patients with a gene called *ATOH1* attached to a harmless viral vector.

Research on the gene Islet1 (*ISL1*) is directed at preventing age-related hearing loss. Normally *ISL1* is involved in development of hair cells but permanently turns itself off after the hair cells are formed in the embryo. When modified *ISL1* was vectored into mouse hair cells, it reprogrammed the hair cells to continuously produce *ISL1*. This reduced age-related hearing loss in the mice.

Stem Cell Therapy

Embryonic stem cells have the capacity of generating all cell types in the body. Unfortunately, when embryonic stem cells are transplanted from a fetus into a patient, they are rejected like any other transplant. To prevent rejection, powerful drugs that partially disable the immune system are required to keep the stem cells alive. Of course disabling the immune system is not a good idea.

Fortunately, the temporary ban on embryonic research led scientists to discover a way to reprogram a patient's own cells into stem cells. Called *induced pluripotent stem cells* (iPS cells), they are transformed from skin or other cells of the patient and function like embryonic stem cells. But since they are taken from the patient herself, iPS cells do not require disabling the immune system. Like new genes used in gene therapy, iPS cells can be delivered to the inner ear by attaching them to inactivated viral vectors.

The inner ear is a prime organ for stem cell therapy because even in deaf patients it remains mostly intact, except for the hair cells. Current research is focusing on producing human hair cells in the laboratory. From there they could be transplanted into the cochlea or be used to identify new drugs to combat deafness. Scientists have already used stem cells to form auditory nerve fibers that can connect to existing hair cells in experimental animals.

Appendix 1
Commonly Used Ear Medications and Their Side Effects

Drug	Drug Class	Use(s)	Side Effects
Acetazolamide	Oral diuretic	Ménière's disease	Tingling in extremities, sun sensitivity
Amoxicillin	Oral antibiotic	Otitis media/externa	Diarrhea, upset stomach
Augmentin	Oral antibiotic	Otitis media/externa	Diarrhea, upset stomach
Betahistine	Oral histamine analogue	Ménière's disease	Headache, upset stomach
Boric Acid	Topical acidifying agent	Otitis externa	Pain/burning, rash, dry skin
Ciprodex	Topical antibiotic	Otitis externa Perforated otitis media	Secondary fungal infection, ear pain
Ciprofloxacin/Levaquin	Oral antibiotic	Otitis media/externa	Diarrhea, upset stomach, rare tendon problem if used with prednisone, muscle/joint ache
Clindamycin	Oral antibiotic	Otitis media/externa, penicillin allergic	Diarrhea (may be severe), upset stomach
Clotrimazole	Topical antifungal	Fungal otitis externa	Itching, burning
Cortisporin	Topical antibiotic	Otitis externa	Potential otoxicity with perforated drum, allergic reaction

Drug	Drug Class	Use(s)	Side Effects
Cresylate	Topical antifungal	Otitis externa	No common side effects reported
CSF-HC Powder	Topical antibiotic and antifungal	Otitis externa	Allergic reaction to sulfa component
Debrox	Topical ceruminolytic	Cerumen impaction	Superinfection, irritation, rash
Depakote (Valproic acid)	Oral	Migrainous vertigo	Drowsiness, diarrhea
Dermotic	Topical steroid	Ear canal eczema	Burning, irritation, dryness, skin thinning
Dyazide/ Hydrochlorothia-zide	Oral diuretic	Ménière's disease	Lightheadedness, sun sensitivity, too much potassium
Floxin	Topical antibiotic	Otitis externa Perforated otitis media	Secondary fungal infection, itching
Prednisone	Oral steroid	Ménière's disease Sudden hearing loss Autoimmune inner ear disease	Short-term use: insomnia, irritability, subtle weight gain, stomach upset, joint problems (rare)
Rocephin (Ceftriaxone)	Injection	Pediatric otitis media	Rash, diarrhea
Valisone	Topical steroid	Ear canal eczema	Burning, irritation, dryness, skin thinning
Vosol	Topical acidifying agent	Otitis externa	Burning, stinging, irritation

Appendix 2
Organizations Offering Information,
Support, and Advocacy

Acoustic Neuroma

Acoustic Neuroma Association
www.anausa.org

American Academy of Otolaryngology—Head and Neck Surgery
(AAO—HNS)
www.entnet.org

National Institutes of Health
http://www.nlm.nih.gov/medlineplus/acousticneuroma.html

Hearing Loss

Alexander Graham Bell Association for the Deaf and Hard of Hearing
(AG Bell)
info@agbell.org

American Academy of Audiology (AAA)
www.audiology.org

American Academy of Otolaryngology—Head and Neck Surgery
(AAO—HNS)
www.entnet.org

American Cochlear Implant Alliance (ACIA)
www.acialliance.org

American Speech-Language-Hearing Association (ASHA)
www.asha.org

Association of Late-Deafened Adults, Inc. (ALDA)
www.alda.org

Gallaudet University
www.clerccenter.gallaudet.edu

Institute for Cochlear Implant Training (ICIT)
CochlearImplantTraining.com

National Association of the Deaf
www.nad.org

National Institute on Deafness and Other Communication Disorders (NIDCD)
www.nidcd.nih.gov

Telecommunications for the Deaf and Hard of Hearing, Inc. (TDI)
www.tdi-online.org

Otitis Media / Otitis Externa

American Academy of Otolaryngology—Head and Neck Surgery (AAO—HNS)
www.entnet.org

American Academy of Pediatrics
www.aap.org

Tinnitus

American Tinnitus Association
www.ata.org

Hyperacusis Network
http://www.hyperacusis.net

Tinnitus Research Initiative
www.tinnitusresearch.org

Vestibular Disorders

Vestibular Disorders Association
www.vestibular.org

Glossary

acoustic neuroma (vestibular schwannoma)—benign tumor that grows on the hearing and balance nerve.

acquired deafness—deafness that is not present at birth.

adenoidectomy—surgical removal of the adenoids.

air conduction—transmission of sound waves through the outer, middle, and inner ear.

Alport syndrome—hereditary condition with hearing loss and kidney disease.

American Sign Language (ASL)—signed language not based on English used by culturally Deaf persons in the United States.

balance—system of equilibrium that allows standing and movement without falling, based on the inner ear, sight, the joints, and touch.

balance disorder—condition that causes loss of equilibrium, dizziness, and vertigo.

barotrauma—injury caused by a change of air pressure as in diving or flying, usually on descent.

central auditory processing disorder (CAPD)—abnormality of the hearing part of the brain that results in difficulty understanding words.

cholesteatoma—benign growth usually caused by repeated middle ear infections; skin cells build up and form a mass that may result in hearing loss, tinnitus, vertigo, and facial nerve weakness; requires surgical removal.

cochlear implant—device that is surgically implanted into the cochlea to provide hearing for people with nerve deafness.

computed tomography scan (CT or CAT scan)—diagnostic imaging procedure based on computer controlled X-rays; shows accurate detail, especially of bone.

conductive hearing loss—hearing loss caused by dysfunction of the outer or middle ear.

decibel (dB)—measurement unit of loudness, based on a logarithmic scale.

decongestant—oral or spray medication that relieves congestion.

dizziness—any sense of lightheadedness, imbalance, or vertigo.

electrocochleography (ECOG)—test of function of the cochlea, measures electrical waves in response to sound.

electronystagmography (ENG)—test of the balance system of the inner ear, records electrical waves caused by eye movements in response to warm and cool stimulation; test has largely been replaced by video recording of eye movement called videonystagmography.

Epley maneuver—treatment of benign paroxysmal positional vertigo (BPPV) based on moving tiny inner ear calcium crystals by head rotation.

Eustachian tube—passageway that connects the middle ear to the nose and throat; opens to equalize pressure in the middle ear.

gene therapy—insertion of new genes to replace defective genes or support function such as hearing.

hair cells—sensory cells of the inner ear topped with hair-like structures (stereocilia), which transform the mechanical energy of sound waves into nerve impulses.

hearing—sound waves are converted to electrical signals that are sent as nerve impulses to the brain where they are heard.

hearing aid—wearable electronic device makes sound louder, includes microphone, amplifier, microprocessor, and receiver.

hyperacusis—phenomenon where normal sounds are perceived to be abnormally loud and painful.

inner ear—part of the ear that includes the cochlea (hearing) and the labyrinth (balance).

labyrinth—balance part of the inner ear, the name reflects a maze of fluid-filled pathways.

labyrinthitis—a viral or bacterial infection of the inner ear; can cause dizziness, loss of balance, and hearing loss.

magnetic resonance imaging (MRI)—diagnostic imaging procedure that uses magnets, radiofrequencies, and a computer to produce detailed images; does not use radiation.

mastoid—hollow bone that forms the posterior portion of the temporal bone behind the ear.

Ménière's disease—inner ear disorder characterized by fluctuating hearing loss, episodes of vertigo, tinnitus, and ear fullness.

meningitis—viral or bacterial infection of the meninges, membranes that surround the brain and the spinal cord.

middle ear—part of the ear that includes the eardrum and three tiny bones of hearing; air filled.

motion sickness—dizziness, sweating, nausea, vomiting caused by motion, as in sea sickness or car sickness.

noise-induced hearing loss—sensorineural hearing loss caused by exposure to loud sound.

nonsyndromic hereditary deafness—inherited hearing loss that is not associated with other inherited abnormalities.

ossicles—middle ear bones including malleus (hammer), incus (anvil), and stapes (stirrup).

otitis externa—inflammation of the outer ear, usually the ear canal as in swimmer's ear.

otitis media—inflammation of the middle ear, usually caused by bacterial infection.

otoacoustic emission (OAE)—test of cochlear function, measurable sound produced by outer hair cells of the cochlea during hearing process.

otolaryngologist—physician/surgeon who specializes in diseases of the ears, nose, throat, and head and neck.

otologist (neurotologist)—sub-specialist in diseases of the ear.

otosclerosis—disorder causing conductive hearing loss, abnormal bone growth that prevents vibration of stapes.

ototoxic drugs—medications that can damage the inner ear.

outer ear—external or outer most portion of the ear including the pinna and ear canal.

perilymph fistula—leakage of inner ear fluid to the middle ear, associated with head trauma, physical exertion, or barotrauma.

postlingually deafened—deafened after having learned language.

prednisone—anti-inflammatory steroid medication used for recent hearing loss, vertigo, or facial paralysis.

prelingually deafened—deafened before learning language.

presbycusis—hearing loss of aging.

round window—opening covered by a membrane that separates the middle ear and inner ear.

sensorineural hearing loss (SNHL)—hearing loss of the cochlea or the hearing nerve, also called nerve deafness.

sign language—language of hand shapes and movements, facial expressions, and body posture.

steroid—medication prescribed to reduce inflammation, hyperactive immune response, and cell death; as in sudden deafness and Bell's palsy.

stroke—sudden loss of consciousness, sensation, or movement caused by blood vessel obstruction or bleeding in the brain.

sudden deafness—hearing loss that occurs within three days, usually of unknown origin but may be caused by Ménière's disease, viral infection, tumors, or ototoxic drugs.

syndromic deafness—congenital deafness associated with other abnormalities such as kidney, eye, or finger anomalies.

tinnitus—sensation of hearing ringing, buzzing, whooshing, etc., without a corresponding external source.

tympanic membrane—eardrum.

tympanometry—ear test of movement of the eardrum, pressure, and volume of the ear canal and middle ear.

tympanoplasty—surgical repair of the eardrum or bones of the middle ear.

vertigo—hallucination of motion, often described as spinning.

vestibular neuronitis—viral infection at the vestibular nerve that causes vertigo.

vestibular schwannoma (acoustic neuroma)—slow-growing benign tumor of the balance nerve that causes hearing loss, tinnitus, and imbalance.

vestibular system—part of the inner ear that is responsible for balance and posture, also called the labyrinth due to its complex structure; abnormalities within it cause vertigo.

References

1 | How the Ear Works

American Academy of Otolaryngology—Head and Neck Surgery. How the ear works. http://www.entnet.org/content/how-ear-works.

Gulya AJ, Minor LB, Poe DS, eds. *Glasscock-Shambaugh Surgery of the Ear*, 6th ed. McGraw-Hill Medical (2010).

2 | Ear Symptoms: What Do They Mean?

Dizziness and Vertigo. U.S. National Library of Medicine. Medline Plus. https://www.nlm.nih.gov/medlineplus/dizzinessandvertigo.html.

Ear Pain. Wikipedia. https://en.wikipedia.org/wiki/Ear_pain.

Hearing Loss. NIH Senior Health. http://nihseniorhealth.gov/hearingloss /hearinglossdefined/01.html.

Tinnitus. National Institute on Deafness and Other Communication Disorders. http://www.nidcd.nih.gov/health/hearing/pages/tinnitus.aspx.

3 | Common Myths about the Ear

Herber RL, King GE, Bent JP. Tympanostomy tubes and water exposure: a practical model. *Arch Otolaryngol Head Neck Surg* (1998); 124: 1118–21.

NIH research portfolio online reporting tools. US Dept. of Health and Human Services. Newborn Hearing Screening. http://report.nih.gov/nihfactsheets /ViewFactSheet.aspx?csid=104 (accessed 4/24/15).

Salata JA, Derkay JS. Water precautions in children with tympanostomy tubes. *Arch Otoloaryngol Head Neck Surg* (1996); 122: 276–80.

Tunkel DE, Bauer CA, Sun GH, Rosenfield RM, et al. Clinical practice guideline: tinnitus. *Otolaryngol Head Neck Surg* (2014); 151: S1–40.

4 | Otitis Media

Agrawal A, Murphy TF. *Haemophilus influenzae* infections in the *H. influenzae Tybe b* conjugate vaccine era. *J Clinical Microbiol* (2011); 49: 3728–32.

Balkany TJ. Acute suppurative otitis media. In: Britton BH, ed. *Common Problems in Otology.* Year Book Medical Publishers (1994).

Balkany TJ, Klein J, Rubin J. Complications of otitis media. In: Balkany TJ, Pashley NRT, eds. *Clinical Pediatric Otolaryngology*. New York: Mosby Co. (1986): 124–32.

Berman S, Balkany TJ, Simmons MA. Otitis media in the neonatal intensive care unit. *Pediatrics* (1978); 62: 198–202.

Elden LM, Coyte PC. Socioeconomic impact of otitis media in North America. *J Otolaryngol* (1998); 27 Suppl 2: 9–16.

Kay DJ, Nelson M, Rosenfeld RM. Meta-analysis of tympanostomy tube sequelae. *Otolaryngol Head Neck Surg* (2001); 124(4): 374.

National Center for Biotechnology Information. http://www.ncbi.nlm.nih.gov /pubmed/19251534.

Tapiainen T, Kujala T, Renko M, et al. Effect of antimicrobial treatment of acute otitis media on the daily disappearance of middle ear effusion: a placebo-controlled trial. *JAMA Pediatr* (2014 July); 168(7): 635–41.

Waseem M. Otitis media treatment and management. *Medscape* (5/25/14). http://emedicine.medscape.com/article/994656-treatment (accessed 7/25/14).

5 | An Overview of Hearing Loss

Blackwell DL, Lucas JW, Clarke TC. Summary health statistics for US adults: National Health Interview Survey, 2012. *Vital Health Stat* (2014); 10: 260.

Gates GA, Mills JH. Presbycusis. *Lancet* (2005); 366: 1111–20.

Lin FR, Niparko JK, Ferrucci L. Hearing loss prevalence in the United States. *Arch Intern Med* (2011 Nov 14); 171(20): 1851–52.

Tadros SF, D'Souza M, Zhu X, Frisina RD. Gene expression changes for antioxidant pathways in the mouse cochlea: relations to age-related hearing deficits. *PLoS One* (2014 Feb 28); 9(2): e90279.

Yoshinaga-Itano C, Sedey AL, Coulter DK, Mehl AL. Language of early- and later-identified children with hearing loss. *Pediatrics* (1998); 102: 1161–71.

6 | Hearing Loss in Children

American Academy of Otolaryngology—Head and Neck Surgery. https:// www.entnet.org/content/genes-and-hearing-loss (accessed 6/7/14).

Maisels MJ, Newman TB. The epidemiology of neonatal hyperbilirubinemia. In: Stevenson DK, Maisels MJ, Watchko JF, eds. *Care of the Jaundiced Neonate*. New York: McGraw-Hill (2012): 97–113.

Mohr PE, Feldman JJ, Dunbar JL, McConkey-Robbins A, Niparko JK, Rittenhouse RK, Skinner MW. The societal costs of severe to profound hearing loss in the United States. *Int J Technol Assess Health Care* (2000 Autumn); 16(4): 1120–35.

National Centers for Disease Control and Prevention. Summary of 2013 EHDI Data. http://www.cdc.gov/ncbddd/hearingloss/2013-data/2013_ehdi_hsfs _summary_a.pdf.

National Institutes of Health. Newborn Hearing Screening. www.report.nih.gov
/nihfactsheets/ViewFactSheet.aspx?csid=104.

Summary of 2009 National CDC EHDI Data. (Revised January 2012). www.cdc
.gov/ncbddd/hearingloss/2009-ata/2009_ehdi_hsfs_summary_508_ok
.pdf.

Toriello HV, Reardon W, Gorlin RJ, eds. *Hereditary Hearing Loss and Its Syn-dromes*. New York: Oxford University Press (2004). www.cdc.gov
/ncbddd/hearingloss/freematerials/parentsguide508.pdf.

7 | Age-Related Hearing Loss

Dai P, Yang W, Jiang S, et al. Correlation of cochlear blood supply with mito-chondrial DNA common deletion in presbycusis. *Acta Otolaryngol* (Mar 2004); 124(2): 130–36.

Gates GA, Mills JH. Presbycusis. *Lancet* (2005); 366: 1111–20.

Pickles JO. Mutation in mitochondrial DNA as a cause of presbycusis. *Audiol Neurootol* (Jan–Feb 2004); 9(1): 23–33.

8 | All about Hearing Aids

Chen DA, Backous DD, Arriaga MA, et al. Phase 1 clinical trial results of the Envoy System: a totally implantable middle ear device for sensorineural hearing loss. *Otolaryngol Head Neck Surg* (2004); 131(6): 904–16.

Fontaine N, Hemar P, Schultz P, Charpiot A, et al. BAHA implant: implantation technique and complications. *Euro Ann of Otorhinolaryngol, Head and Neck Diseases* (2014); 131(1): 69–74. www.nidcd.nih.gov/health/statistics/pages.

Kiringoda R, Lustig LR. A meta-analysis of the complications associated with osseointegrated hearing aids. *Otology & Neurotology* (2013); 34: 790–94.

Kraus EM, Shohet JA, Catalano PJ. Envoy Esteem totally implantable hearing system: Phase 2 trial, 1-year hearing results. *Otolaryngol Head Neck Surg* (2011); 145(1): 100–109.

9 | All about Cochlear Implants

Balkany TJ, Hodges AV, Eshraghi AA, Butts S, et al. Cochlear implants in chil-dren—a review. *Acta Otolaryngol* (2002); 122: 356–62.

Balkany TJ, Hodges AV, Goodman KW. Ethics of cochlear implantation in young children. *Otolaryngol Head Neck Surg* (1996); 114: 748–55.

Bertling T. *A Child Sacrificed to the Deaf Culture*. Kodiak Media Group (1994).

Eter EG, Balkany TJ. Pediatric cochlear implant surgery. *Otolaryngol Head Neck Surg* (2009); 20(3): 202–5.

Gates GA, Anderson ML, McCurry SM, Feeney MP, et al. Central auditory dys-function as a harbinger of Alzheimer dementia. *Arch Otolaryngol Head Neck Surg* (2011 Apr); 137(4): 390–95.

National Institutes of Health. http://www.nidcd.nih.gov/health/hearing/pages
/coch.aspx (accessed 10/19/14).

Oulade OA, Koo DS, LaSasso CJ, Eden GF. Neuroanatomical profiles of deafness in the context of native language experience. *J Neurosci* (2014 Apr 16); 34(16): 5613–20.

Ruckenstein MJ, Hedgepeth C, Rafter KO, et al. Tinnitus suppression in patients with cochlear implants. *Otology & Neurotology* (2001); 22: 200–204.

Summerfield AQ, Cirstea SE, Roberts KL, et al. Incidence of meningitis and of death from all causes among users of cochlear implants in the United Kingdom. *J Public Health* (2005); 27(1): 55–61.

10 | Dizziness and Vertigo

Balkany TJ, Finkel RS. The dizzy child. *Ear and Hearing* (1986); 7: 138–42.

Balkany TJ, Sires B, Arenberg IK. Bilateral aspects of Meniere's disease. *Otolaryngol Clin N Am* (1980); 13: 603–9.

Baloh RW, Honrubia V, Jacobson K. Benign positional vertigo. *Neurology* (1987); 37: 371–78.

Neuhauser HK. Epidemiology of vertigo. *Current Opinion in Neurology* (2007); 20: 40–46.

Rine RM, Schubert MC, Balkany TJ. Visual-vestibular habituation and balance training for motion sickness. *Physical Therapy* (1999); 79: 949–57.

Strupp M, Zingler VC, Arbusow V, Niklas D, Maag KP, Dieterich M, et al. Methylprednisolone, valacyclovir, or the combination for vestibular neuritis. *N Engl J Med* (2004 Jul 22); 351(4): 354–61.

Tarantino S, Capuano A, Torriero R, Citti M, et al. Migraine equivalents as part of migraine syndrome in childhood. *Pediatr Neurol* (2014); 51(5): 645–49.

Weber PC, Bluestone CD, Perez B. Outcome of hearing and vertigo after surgery for congenital perilymphatic fistula in children. *Am J Otolaryngol* (2003); 24(3): 138–42.

11 | Tinnitus: Ringing in the Ears

Folmer RL, Griest SE. Tinnitus and insomnia. *Am J Otolaryngol* (2000); 21: 285–93.

Kochkin S, Tyler R, Born J. MarkeTrak VIII: The prevalence of tinnitus in the United States and the self-reported efficacy of various treatments. *Hearing Review* (2011); 18(12): 10–26.

Koizumi T, Nishimura T, Yamashita A, Yamanaka T, Imamura T, Hosoi H. Residual inhibition of tinnitus induced by 30-kHz bone-conducted ultrasound. *Hear Res* (2014 Feb 11): pii: S0378-5955(14)00019-7.

Ruckenstein MJ, Hedgepeth C, Rafter KO, et al. Tinnitus suppression in patients with cochlear implants. *Otology & Neurotology* (2001); 22: 200–204.

Tunkel DE, Bauer CA, Sun GH, Rosenfield RM, et al. Clinical practice guideline: tinnitus. *Otolaryngol Head Neck Surg* (2014); 151: S1–40.

Vanneste S, Fregni F, De Ridder D. Head-to-head comparison of transcranial random noise stimulation, transcranial AC stimulation, and transcranial DC stimulation for tinnitus. *Front Psychiatry* (2013 Dec 18); 4: 158–62.

Vogler DP, Robertson D, Mulders WH. Hyperactivity following unilateral hearing loss in characterized cells in the inferior colliculus. *Neuroscience* (2014 Jan 24): pii: S0306–4522(14)00028-1.

12 | Swimmer's Ear

Balkany TJ, Ress BD. Infections of the external ear. In: Cummings CW, Fredrickson JM, Harker LA, Krause CJ, Richardson MA, Schuller DE, eds. *Otolaryngology Head and Neck Surgery*, 3rd ed. St Louis: Mosby–Year Book, Inc. (1998); 4: 2979–86.

Carfrae MJ, Kesser BW. Malignant otitis externa. *Otolaryngol Clin North Am* (2008); 41: 537–49.

Roland PS, Stroman DW. Microbiology of acute otitis externa. *Laryngoscope* (2002); 112: 1166–77.

13 | Ear Wax and Foreign Bodies

Kavanagh K, Litovitz T. Miniature battery foreign bodies in auditory and nasal cavities. *JAMA* (1986); 255(11): 1470–72.

Seely DR, Quigley SM, Langman AW. Ear candles—efficacy and safety. *Laryngoscope* (1996); 106(10): 1226–29.

Thompson SM, Wein RO, Dutcher PO. External auditory canal foreign body removal: management practices and outcomes. *Laryngoscope* (2003); 113(11): 1912–15.

14 | Malformations of the Outer Ear

Brackmann DE, Shelton C, Arriaga MA. *Otologic Surgery*. Philadelphia, PA: Saunders/Elsevier (2010).

Conway H, Wagner K. Congenital anomalies of the head and neck. *Plast Reconstr Surg* (1965); 36: 71–79.

Mustarde JC. The correction of prominent ears using simple mattress sutures. *Br J Plast Surg* (1963); 16: 170–78.

Nazarian R, Eshraghi AA. Otoplasty for the protruded ear. *Semin Plast Surg* (2011); 25(4): 288–94.

Sivayoham E, Woolford TJ. Current opinion on auricular reconstruction. *Curr Opin Otolaryngol Head Neck Surg* (2012); 20(4): 287–90.

15 | Bony Growths of the Ear Canal

House JW, Wilkinson EP. External auditory exostoses: evaluation and treatment. *Otolaryngol Head Neck Surg* (2008); 138(5): 672–78.

King JF, Kinney AC, Iacobellis SF 2nd, Alexander TH, Harris JP, Torre P 3rd, Doherty JK, Nguyen QT. Laterality of exostosis in surfers due to evaporative cooling effect. *Otol Neurotol* (2010); 31(2): 345–51.

Wong BJ, Cervantes W, Doyle KJ, Karamzadeh AM, Boys P, Brauel G, Mushtaq E. Prevalence of external auditory canal exostoses in surfers. *Arch Otolaryngol Head Neck Surg* (1999); 125(9): 969–72.

16 | Cancer of the Outer Ear

Alam M, Ratner D. Cutaneous squamous-cell carcinoma. *N Engl J Med* (2001 Mar); 344(13): 975–83.

Cole MD, Jakowatz J, Evans GR. Evaluation of nodal patterns for melanoma of the ear. *Plast Reconstr Surg* (2003 Jul); 112(1): 50–56.

Devaney KO, Boschman CR, Willard SC, Ferlito A, Rinaldo A. Tumours of the external ear and temporal bone. *Lancet Oncology* (2005 Jun); 6(6): 411–20.

Mohs F, Larson P, Iriondo M. Micrographic surgery for the microscopically controlled excision of carcinoma of the external ear. *J Am Acad Dermatol* (1988 Oct); 19(4): 729–37.

Moody SA, Hirsch BE, Myers EN. Squamous cell carcinoma of the external auditory canal: an evaluation of a staging system. *Am J Otol* (2000 Jul); 21(4): 582–88.

Witmanowski H, Lewandowicz E, Sobieszek D, Rykala J, Luczkowska M. Facial skin cancers: general information and an overview of treatment methods. *Postep Derm Alergol* (2012); 29(4): 240–55.

17 | Trauma: Cauliflower Ear

DeSanti L. Pathophysiology and current management of burn injury. *Advances in Skin & Wound Care* (2005 Jul–Aug); 18(6): 323–32; quiz 32–34.

Greywoode JD, Pribitkin EA, Krein H. Management of auricular hematoma and the cauliflower ear. *Facial Plastic Surgery* (2010 Dec); 26(6): 451–55.

Mudry A, Pirsig W. Auricular hematoma and cauliflower deformation of the ear: from art to medicine. *Otol Neurotol* (2009 Jan); 30(1): 116–20.

Petrone P, Kuncir EJ, Asensio JA. Surgical management and strategies in the treatment of hypothermia and cold injury. *Emergency Medicine Clinics of North America* (2003 Nov); 21(4): 1165–78.

Pham TV, Early SV, Park SS. Surgery of the auricle. *Facial Plastic Surgery* (2003 Feb); 19(1): 53–74.

18 | Perforated Eardrum and Tympanoplasty

Casselbrant ML, Kaleida PH, Rockette HE, et al. Efficacy of antimicrobial prophylaxis and of tympanostomy tube insertion for prevention of recurrent acute otitis media: results of a randomized clinical trial. *Pediatr Infect Dis J* (1992 Apr); 11(4): 278–86.

Hahn Y, Bojrab DI. Outcomes following ossicular chain reconstruction. *Ear Nose Throat J* (2013); 92: 250–54.

Hardman J, Muzaffar J, Nankivell P, Coulson C. Tympanoplasty for chronic tympanic membrane perforation in children: systematic review and meta-analysis. *Otol Neurotol* (2015 Jun); 36(5): 796–804.

Neumann A, Kevenhoerster K, Gostian AO. Long-term results of palisade cartilage tympanoplasty. *Otol Neurotol* (2010 Aug); 31(6): 936–39.

Subotic R, Mladina R, Risavi R. Congenital bony fixation of the malleus. *Acta Otolaryngol* (1998); 118: 833–36.

Tang S, Brown KD. Success of lateral graft technique for closure of tympanic membrane perforations. *Otol Neurotol* (2015 Feb); 36(2): 250–53.

Yilmaz MS, Guven M, Kayabasoglu G, Varli AF. Comparison of the anatomic and hearing outcomes of cartilage type 1 tympanoplasty in pediatric and adult patients. *Eur Arch Otorhinolaryngol* (2015 Mar); 272(3): 557–62.

19 | Otosclerosis and Stapedotomy

Cruise AS, Singh A, Quiney RE. Sodium fluoride in otosclerosis treatment: review. *J Laryngol Otol* (2010 Jun); 124(6): 584–86.

Declau F, van Spaendonck M, Timmermans JP, et al. Prevalence of histologic otosclerosis: an unbiased temporal bone study in Caucasians. *Adv Otorhinolaryngol* (2007); 65: 6–16.

Gristwood RE, Venables WN. Pregnancy and otosclerosis. *Clin Otolaryngol and Allied Sciences* (2007); 8: 2005–10.

McManus LJ, Stringer MD, Dawes PJ. Iatrogenic injury of the chorda tympani: a systematic review. *J Laryngol Otol* (2012); 126(1): 8–14.

Stucken EZ, Brown KD, Selesnick SH. The use of KTP laser in revision stapedectomy. *Otol Neurotol* (2012); 33(8): 1297–99.

Vincent R, Bittermann AJ, Oates J, Sperling N, Grolman W. KTP versus CO2 laser fiber stapedotomy for primary otosclerosis: results of a new comparative series with the otology-neurotology database. *Otol Neurotol* (2012); 33(6): 928–33.

20 | Mastoiditis and Cholesteatoma

Jung JY, Lee DH, Wang EW, Nason R, Sinnwell TM, Vogel JP, Chole RA. *P. aeruginosa* infection increases morbidity in experimental cholesteatomas. *Laryngoscope* (2011); 121(11): 2449–54.

Pritchett CV, Thorne MC. Incidence of pediatric acute mastoiditis: 1997–2006. *Arch Otolaryngol Head Neck Surg* (2012); 138: 451–55.

Sadé J, Ar A. Middle ear and auditory tube: middle ear clearance, gas exchange, and pressure regulation. *Otolaryngol Head Neck Surg* (1997); 116(4): 499–524.

Taylor MF, Berkowitz RG. Indications for mastoidectomy in acute mastoiditis in children. *Ann Otol Rhinol Laryngol* (2004); 113(1): 69–72.

Teele DW, Klein JO, Rosner B. Epidemiology of otitis media during the first seven years of life in children in greater Boston: a prospective, cohort study. *J Infect Dis* (1989); 160(1): 83–94.

Vartiainen E. Factors associated with recurrence of cholesteatoma. *J Laryngol Otol* (1995); 109(7): 590–92.

21 | Ménière's Disease

Ahsan SF, Standring R, Wang Y. Systematic review and meta-analysis of Meniett therapy for Ménière's disease. *Laryngoscope* (2015); 125: 203–8.

Balkany TJ, Pillsbury HC, Arenberg IK. Defining and qualifying Ménière's disease. *Otolaryngol Clin N Am* (1980); 13: 589–96.

Balkany TJ, Sires B, Arenberg IK. Bilateral aspects of Ménière's disease. *Otolaryngol Clin N Am* (1980); 13: 603–9.

Basel T, Lutkenhoner B. Auditory threshold shifts after glycerol administration to patients with suspected Ménière's disease: a retrospective analysis. *Ear Hear* (2013 May–Jun); 34(3): 370–84.

Gabra N, Saliba I. The effect of intratympanic methylpredisilone and gentamicin injection on Ménière's disease. *Otolaryngol Head Neck Surg* (2013); 148: 642–47.

Monsell EM, Balkany TJ, Gates GA, et al. Committee on Hearing and Equilibrium guidelines for the diagnosis and evaluation of therapy in Ménière's disease. *Otolaryngol Head Neck Surg* (1995); 113: 181–85.

National Institute on Deafness and Other Communication Disorders. Ménière's disease. http://www.nidcd.nih.gov/health/balance/. See more at: http://vestibular.org/Ménières-disease#sthash.923MwWrr.dpuf.

Pullens B, van Benthem PP. Intratympanic gentamicin for Ménière's disease or syndrome. *Cochrane Database Syst Rev* (2011); 3: CD008234.

Salt AN, Plontke SK. Endolymphatic hydrops: pathophysiology and experimental models. *Otolaryngol Clin North Am* (2010); 43: 971–83.

Strupp M, Hupert D, et al. Long-term prophylactic treatment of attacks of vertigo in Ménière's disease—comparison of a high with a low dosage of betahistine in an open trial. *Acta Otolaryngol* (2008); 128(5): 520–24.

22 | Noise-Induced Hearing Loss

Balkany TJ, Eshraghi AA, Jiao H, Polak M, et al. Mild hypothermia protects auditory function during cochlear implant surgery. *Laryngoscope* (2005); 115: 1543–47.

Hamernik RP, Qiu W, Davis B. Cochlear toughening, protection, and potentiation of noise-induced trauma by non-Gaussian noise. *J Acoust Soc Am* (2003); 113(2): 969–76.

Mahmood G, Mei Z, Hojjat H, Pace E, Kallakuri S, Zhang JS. Therapeutic effect of sildenafil on blast-induced tinnitus and auditory impairment. *Neuroscience* (2014); 269: 367–82.

https://www.nidcd.nih.gov/health/hearing/pages/noise.aspx (accessed 5/27/14).

https://www.osha.gov/SLTC/noisehearingconservation (accessed 7/24/14).

Rajan R. Cochlear outer-hair-cell efferents and complex-sound-induced hearing loss: protective and opposing effects. *J Neurophysiol* (2001); 86: 3073–76.

Rask-Andersen H, Ekvall L, Scholtz A, Schrott-Fischer A. Structural/audiometric correlations in a human inner ear with noise-induced hearing loss. *Hear Res* (2000); 141(1–2): 129–39.

Shargorodsky J, Curhan SG, Curhan GC, Eavey R. Change in prevalence of hearing loss in US adolescents. *JAMA* (2010); 304: 772–78.

Tahera Y, Meltser I, Johansson P, Bian Z, et al. NF-kappaB mediated glucocorticoid response in the inner ear after acoustic trauma. *J Neurosci Res* (2006); 83: 1066–76.

Tahera Y, Meltser I, Johansson P, Salman H, et al. Sound conditioning protects hearing by activating the hypothalamic-pituitary-adrenal axis. *Neurobiol Dis* (2007 Jan); 25(1): 189–97.

23 | Sudden Deafness and Autoimmune Disease

Alexander TH, Weisman MH, Derebery JM, et al. Safety of high-dose corticosteroids for the treatment of autoimmune inner ear disease. *Otology & Neurotology* (2009); 30: 443–48.

Awad Z, Huins C, Pothier DD. Antivirals for idiopathic sudden sensorineural hearing loss. *Cochrane Database Syst Rev* (2012 Aug 15); 8: CD006987.

Battaglia A, Burchette R, Cueva R. Combination therapy (intratympanic dexamethasone + high-dose prednisone taper) for the treatment of idiopathic sudden sensorineural hearing loss. *Otol & Neurotol* (2008); 29: 453–60.

Bovo R, Ciorba A, Martini A. The diagnosis of autoimmune inner ear disease: evidence and critical pitfalls. *Eur Arch Otorhinolaryngol* (2009); 266: 37–40.

Chays A, Dubreuil C, Vaneecloo FM, Magnan J. Sudden deafness and neurinoma. *Eur Ann Otorhinolaryngol Head Neck Dis* (2011 Jan); 128(1): 24–29.

Liu SC, Kang BH, Lee JC, et al. Comparison of therapeutic results in sudden sensorineural hearing loss with/without additional hyperbaric oxygen therapy: a retrospective review of 465 audiologically controlled cases. *Clin Otolaryngol* (2011 Apr); 36(2): 121–28.

Mijovic T, Zeitouni A, Colmegna I. Autoimmune sensorineural hearing loss: the otology-rheumatology interface. *Rheumatology (Oxford)* (2013 May); 52(5): 780–89.

Ryan AF, Harris JP, Keithley EM. Immune-mediated hearing loss: basic mechanisms and options for therapy. *Acta Otolaryngol Suppl* (2002); 548: 38–43.

Sara SA, Teh BM, Friedland P. Bilateral sudden sensorineural hearing loss: review. *J Laryngol Otol* (2014 Jan); 128 Suppl 1: S8–15.

Shaia FT, Sheehy JL. Sudden sensori-neural hearing impairment: a report of 1,220 cases. *Laryngoscope* (1976 Mar); 86(3): 389–98.

Wei BP, Mubiru S, O'Leary S. Steroids for idiopathic sudden sensorineural hearing loss. *Cochrane Database Syst Rev* (2006 Jan 25); 1: CD003998.

Wu X, Chen K, Sun L, Yang Z, et al. Magnetic resonance imaging–detected inner ear hemorrhage as a potential cause of sudden sensorineural hearing loss. *Am J Otolaryngol* (2014 May–Jun); 35(3): 318–23.

24 | Ototoxic Drugs

Amarasena IU, Chatterjee S, Walters JA, Wood-Baker R, Fong KM. Platinum versus non-platinum chemotherapy regimens for small cell lung cancer. *Cochrane Database Syst Rev (2015 Aug); 2 (8): CD006849.* Cochrane Library. Published Online: 2 AUG 2015 doi: 10.1002/14651858.CD006849.pub3.

Bas E, Van De Water TR, Gupta C, Dinh J, et al. Efficacy of three drugs for protecting against gentamicin-induced hair cell and hearing losses. *Br J Pharmacol* (2012 Jul); 166(6): 1888–904.

Bitner-Glindzicz M, Rahman S. Ototoxicity caused by aminoglycosides is severe and permanent in genetically susceptible people. *British Med J* (2007); 335: 784.

Ojano-Dirain CP, Antonelli PJ, Le Prell CG. Mitochondria-targeted antioxidant MitoQ reduces gentamicin-induced ototoxicity. *Otol Neurotol* (2014 Mar); 35(3): 533–39.

25 | Acoustic Neuroma and Other Tumors

Acoustic Neuroma Association (2013 Nov). *Acoustic Neuroma Basic Overview.* ANA Patient Information Booklets: 3.

Forbes JA, Brock AA, Ghiassi M, Thompson RC, Haynes DS, Tsai BS. Jugulotympanic paragangliomas: 75 years of evolution in understanding. *Neurosurg Focus* (2012); 33(2): E13.

Hasegawa T, Kida Y, Kobayashi T, Yoshimoto M, et al. Long-term outcomes in patients with vestibular schwannomas treated using gamma knife surgery: 10-year follow up. *J Neurosurg* (2013 Dec); 119 Suppl: 10–16.

Herscovici Z, et al. Natural history of conservatively treated meningiomas. *Neurology* (2004 Sep 28); 63(6): 1133–34.

26 | Hearing and Balance Tests

Balkany TJ, Lonsbury-Martin BL. Otoacoustic emissions. *Am J Otol* (Monograph) (1994); 15(1) Suppl: 1–38.

Balkany TJ, Zarnock MJ. Impedance tympanometry in infants. *Audiology and Hearing Education* (1978); 4: 17–19.

Burkard RF, Don M, Eggermont JJ. *Auditory Evoked Potentials: Basic Principles and Clinical Application.* Hagerstown, MD: Lippincott Williams & Wilkins (2007).

Gates GA, Anderson ML, McCurry SM, Feeney MP, Larson EB. Central auditory dysfunction as a harbinger of Alzheimer dementia. *Arch Otolaryngol Head Neck Surg* (2011 Apr); 137(4): 390–95.

genome.gov/www.nchpeg.org/shla/site.html (accessed 7/20/14).

Musiek FE, Chermak GD. *Handbook of Central Auditory Processing Disorder.* Volume 1: *Auditory Neuroscience and Diagnosis.* San Diego, CA: Plural Publishing (2007): 448.

Roush J, Grose J. Principles of audiometry. In: Van De Water TR, Staecker H, eds. *Otolaryngology: Basic Science and Clinical Review.* New York: Thieme (2006): 374–84.

27 | The Ear and Scuba Diving

Bove AA. *Bove and Davis' Diving Medicine*, 4th ed. Philadelphia, PA: Saunders/ Elsevier (2003).

Lynch JH, Bove AA. Diving medicine: a review of current evidence. *J Am Board Fam Med* (2009); 22(4): 399–407.

28 | Airplane Ear

Affleck J, Angelici A, Baker S, Brook T, et al. Cabin cruising altitudes for regular transport aircraft. *Aviat Space Environ Med* (2008 Apr); 79(4): 433–39.

Mitchell-Innes A, Young E, Vasiljevic A, Rashid M. Air travelers' awareness of the preventability of otic barotrauma. *Laryngol Otol* (2014 Jun); 128(6): 494–98.

29 | Bell's Palsy and the Facial Nerve

Abdel-Aziz M, Azab NA, Khalifa B, Rashed M, Naguib N. The association of Varicella zoster virus reactivation with Bell's palsy in children. *Int J Pediatr Otorhinolaryngol* (2015 Mar); 79(3): 328–31.

Baugh RF, Basura GJ, Ishii LE, Schwartz SR, et al. Clinical practice guideline: Bell's palsy executive summary. *Otolaryngol Head Neck Surg* (2013 Nov); 149(5): 656–63.

Gronseth GS, Paduga R. Evidence-based guideline update: steroids and antivirals for Bell palsy: report of the Guideline Development Subcommittee of the American Academy of Neurology. *Neurology* (2012 Nov 27); 79(22): 2209–13.

Morris AM, Deeks SL, Hill MD, et al. Annualized incidence and spectrum of illness from an outbreak investigation of Bell's palsy. *Neuroepidemiology* (2002); 21(5): 255–61.

Peitersen E. The natural history of Bell's palsy. *Am J Otol* (1982); 4(2): 107–11.

Sullivan FM, Swan IR, Donnan PT, et al. Early treatment with prednisolone or acyclovir in Bell's palsy. *N Engl J Med* (October 2007); 357(16): 1598–607.

Peng T, Dong Y, Zhu G, Xie D. Induced pluripotent stem cells: landscape for studying and treating hereditary hearing loss. *J Otology:* doi:10.1016/j.joto .2015.02.001.

Pinyon JL, Tadros SF, Froud KE, et al. Close-field electroporation gene delivery using the cochlear implant electrode array enhances the bionic ear. *Sci Transl Med* (2014); 6(233): 233ra54.

Safety, Tolerability and Efficacy for CGF166 in Patients with Bilateral Severe-to-profound Hearing Loss. https://clinicaltrials.gov/ct2/show/NCT02132130 (accessed 9/27/16).

Sun Y, Dykes IM, Liang X, et al. A central role for Islet1 in sensory neuron development linking sensory and spinal gene regulatory programs. *Nat Neurosci* (2008); 11: 1283–93.

Index

CT. *See* computed tomography (CT)

goiter, 41, 43, 191
Goldenhar syndrome, 107
gravity sensation, 3, 6, 7, 73
GT. *See* glomus tumor (GT)

Haemophilus influenzae, 25, 26, 68, 138
hair cells, 6, 7, 73, 218; drug-induced damage, 167; gene therapy research, 210; noise-induced damage, 158; presbycusis and, 52; sensorineural hearing loss and, 32; stem cell research, 211
hammer. *See* malleus
HBO (hyperbaric oxygen therapy), 164
headache: acoustic neuroma and, 171, 175; in acute otitis media, 23; betahistine-induced, 213; caffeine-withdrawal, 151; glomus tumor and, 177; in meningitis, 69; migrainous vertigo, 77
head trauma, 9, 51, 77, 84, 132, 219
hearing, 6, 218
hearing aids, 55–60, 218; for age-related hearing loss, 50, 53, 54; vs. assistive listening devices, 58–59; bone-anchored, 59–60, 110–11; cerumen impaction with, 104; changing batteries, 56–57; for children, 16, 35, 49, 61; cholesteatoma and, 132; vs. cochlear implant, 62; for conductive hearing loss, 34, 110–11; contact dermatitis with, 100; cost, 16, 55, 58, 134, 135; fungal infection with, 95; hearing threshold and, 35, 65–66; how they work, 55; in Ménière's disease, 148; middle ear implants, 60; myths about, 16; for otosclerosis, 34, 134, 135; purchase, 57–58; for sensorineural hearing loss, 15; size and position in ear, 56; for tinnitus, 15, 88; tip left in ear after removal of, 105; who should try, 57
hearing loss, 8–9, 31–38; age-related, 36, 50–54, 219 (*see also* presbycusis); causes in adults, 31; conductive, 32, 33–34, 217; due to ototoxic drugs, 31, 44, 46t, 54, 153, 167–70, 219; incidence, 31; labyrinthitis and, 75, 76; in Ménière's disease, 31, 32, 74, 162, 163; mixed, 32; noise-induced, 15, 31, 36, 156–60, 219; organizations for, 215–16; otitis media and, 21, 25, 29; otosclerosis and, 31, 34, 134; self-evaluation for, 37–38; sensorineural, 32–33, 183, 219; severity (threshold) of, 35;

speech recognition and, 35–36, 38; sudden deafness, 31, 161–64, 220; tinnitus and, 82; treatment, 32
hearing loss in children, 31, 39–49; causes, 36; congenital, 37, 39–45 (*see also* congenital hearing loss); diagnosis in infants, 14, 45–46; evaluation in infants, 48; genetic testing, 42, 190–91; hearing aids for, 16, 35, 49, 61; high-risk factors in newborns, 46–48; interacting with doctor, 49; language development and, 14, 16, 24, 29, 31, 35, 36–37, 39, 47, 66; newborn screening, 14, 39, 45–47; normal hearing and language milestones, 47; severity (threshold) of, 35
hearing tests, 35, 183–87; air and bone conduction, 183–84; brainstem auditory response, 45–46, 185, 187; in infants, 48; in Ménière's disease, 150; otoacoustic emissions, 45–46, 186–87, 219; during ototoxic drug use, 168–69; pure tone audiometry, 184–85; self-administered questionnaires, 37–38; speech audiometry, 185–86; for tinnitus, 81, 84, 86; tympanometry, 48, 186, 220
hearing threshold: hearing aids and, 35, 65–66; pure tone audiometry, 184–85; severity of hearing loss, 35; temporary and permanent threshold shifts, 158
hearing your breathing, 11
hearing your own voice, 11, 33–34
hematoma: auricular, 120–21; after otoplasty, 108
herpes viruses: antiviral drugs for, 164, 207; Bell's palsy and, 205, 207; congenital deafness and, 43; sudden deafness and, 164
HIV disease, 13, 43, 101
hydrocephalus, 9, 30, 174
hydrochlorothiazide, 214
hyperacusis, 218; in Lyme disease, 190; in Ménière's disease, 149; tinnitus and, 90
Hyperacusis Network, 216
hyperbaric oxygen therapy (HBO), 164
hyperbilirubinemia, 44–45, 46
hyperventilation, 79
hypoglycemia, 10, 73, 79
hypotension, postural, 79–80
hypothermia, for noise-induced hearing loss, 160

childhood, 78; in Ménière's disease, 147; video nystagmography, 150, 188–89, 218

nystatin, 98

OAE (otoacoustic emission) test, 45–46, 48, 186–87, 219

occlusion effect, 11, 33–34

Occupational Safety and Health Administration (OSHA), 157

ofloxacin (Floxin Otic), 98, 214

OM. See otitis media (OM)

OME (otitis media with effusion), 23–24, 27

orthostatic hypotension, 79–80

OSHA (Occupational Safety and Health Administration), 157

ossicles, 5, 7, 219; cholesteatoma affecting, 132, 140; congenital fixation, 133; hearing loss due to damage of, 34; infection-related erosion of, 131–32; otosclerosis, 134–36, 219; prostheses, 132; traumatic dislocation, 132–33. See also incus; malleus; stapes

osteitis, 100

osteomas, 112, 113–14

osteomyelitis, skull base, 101

otitis externa, 93–101, 219; anatomy of infection, 93–94; bacterial, 94; causes, 93; cellulitis and, 99; chronic dermatitis and, 99–100; in diabetics or immunosuppressed persons, 100–101; fungal, 95; hearing loss due to, 34; necrotizing, 101; organizations for, 216; prevention, 95–97; referral to ear specialist, 99; side effects of medications for, 213–14; spread of infection, 100; treatment, 97–99

otitis media (OM), 10, 11, 21–30, 219; acute, 23, 25–27; adenoidectomy for, 28–29; adhesive, 29; causes, 22–23; cholesteatoma and, 139; chronic suppurative, 29; complications, 29–30; diagnosis, 23, 24–25; due to Eustachian tube dysfunction, 22; eardrum perforation due to, 29, 128; ear tubes for, 27–28; with effusion, 23–24, 27; hearing loss due to, 21, 25, 34; mastoiditis and, 137; neonatal, 45; organizations for, 216; prevalence, 21–22; prevention, 25; side effects of medications for, 124, 213; spread of infection, 21, 29–30; treatment, 25–29

otoacoustic emission (OAE) test, 45–46, 48, 186–87, 219

otolaryngologist, 219

otologist, 219

otoplasty, for microtia, 108

otosclerosis, 134–36, 219; hearing loss due to, 31, 34, 134; treatment, 34, 134–36

ototoxicity, drug-induced, 31, 46t, 54, 153, 167–70, 219; congenital deafness and, 44; drugs associated with, 167–68, 169; genetic defect and, 168; precautions and monitoring for, 168–70; sudden deafness due to, 220; symptoms, 168. See also specific drugs

outer ear, 3–4, 219

outer ear disorders: bony growths of ear canal, 112–14; cancer, 115–19; conductive hearing loss due to, 33–34; ear wax impaction, 102–5; foreign bodies in ear canal, 105–6; malformations, 34, 107–11; otitis externa (swimmer's ear), 93–101

oxaplatin ototoxicity, 44

oxycodone, 194

pain, 10; acoustic neuroma, 173; adenoidectomy, 28; airplane ear, 199, 200, 202; bacterial labyrinthitis, 6; barotrauma, 128, 194, 195, 197; Bell's palsy, 206, 207; burns, 122; cellulitis, 99; cochlear implant infection, 70; drug-induced, 213t; ear, 85, 101; eardrum perforation, 11, 13, 128; frostbite, 122; glandular tumors of ear canal, 119; from loud noise, 15, 218; mastoiditis, 137; meningitis, 69; otitis externa, 94, 98; otitis media, 23; referred, 10, 14; skull base osteomyelitis, 101; swimmer's ear, 11; temporomandibular joint, 10, 14; tympanoplasty, 131

paraganglioma. See glomus tumor (GT)

partial ossicular replacement prosthesis (PORP), 132

peanut ear, 108

Pendred syndrome, 41, 43

perichondritis, 100

perichondrium, 120–21

perilymph fistula (PLF), 74, 77, 197–98, 203, 219

permanent threshold shift (PTS), 158

petrositis, 30

pinna, 3, 7, 94, 219; auricular hematoma, 120–21; basal cell carcinoma, 115; burns, 122–23; contact dermatitis, 100; frostbite, 122;

temporal bone: basal cell carcinoma, 115; glandular tumors, 119; squamous cell carcinoma, 117–18

temporary threshold shift (TTS), 158

temporomandibular joint (TMJ) pain, 10, 14

thermal injuries, 122–23

thyroid disorders, 10, 74, 145, 151, 191

thyroid function studies, 150

tinnitus, 8, 9, 12, 13, 81–90, 220; acoustic neuroma and, 76, 171, 172; airplane ear and, 203; autoimmune inner ear disease and, 165; Bell's palsy and, 206; causes, 81–82; cochlear implants and, 70, 89; doctor's evaluation of, 84–86; drug-induced, 167, 168, 169; glomus tumor and, 177, 178; health food supplements for, 15; hearing loss and, 82; hearing tests for, 86; history taking, 84–85; hyperacusis and, 90; loudness of, 84; in Lyme disease, 190; measurement, 82; in Ménière's disease, 10, 74, 145, 147, 148–49; meningioma and, 76; myoclonus and, 83–84; objective, 82–84, 89–90; in one ear, 82; organizations for, 216; perilymph fistula and, 77, 198; physical examination, 85; prevalence, 82; research, 90, 216; subjective, 81, 82, 86–89; sudden deafness and, 161; treatment, 15, 81, 86–90; vascular (pulse synchronous), 82–83; in viral labyrinthitis, 75

Tinnitus Research Initiative, 216

TM (tympanic membrane), 127, 220. See also eardrum

TMJ (temporomandibular joint) pain, 10, 14

tobramycin ototoxicity, 44, 168

tonsillectomy, 28–29

TORP (total ossicular replacement prosthesis), 132

total ossicular replacement prosthesis (TORP), 132

toxoplasmosis, 43

Toynbee maneuver, 196

trauma: auricular hematoma, 120–21; barotrauma, 128, 196–97, 217; blast injuries, 158; burns, 122–23; dislocation of hearing bones, 132–33; eardrum perforation due to, 128; frostbite, 122; head, 9, 51, 77, 84, 132, 219; lacerations, 121–22; perilymph fistula due to, 77, 198, 219

trazodone, 87

Treacher Collins syndrome, 107

TTS (temporary threshold shift), 158

Tumarkin's otologic crisis, 146

tumors, 9; acoustic neuroma, 76–77, 171–76; benign, of auditory nerve, 31, 32; glomus tumor, 177–79; meningioma, 76–77, 176–77; middle ear, 6; tinnitus due to, 9, 15

tympanic membrane (TM), 127, 220. See also eardrum

tympanomastoidectomy, 140

tympanometry, 24, 186, 220

tympanoplasty, 127, 128, 129–31, 132, 220; care after, 131; graft material for, 130–31; with middle ear reconstruction for conductive hearing loss, 131–33; outcome, 131; techniques, 129–30

tympanosclerosis, 132

tympanostomy tubes. See ear tubes

urea dehydration test, 150

utricle, 6–7, 73, 75

vaccines: Haemophilus influenzae, 25, 26, 68; Meningococcus, 68; pneumococcus, 25, 26, 68

Valisone, 214

Valium (diazepam), 152

valproic acid, 214

vancomycin ototoxicity, 169

vascular malformations, 83, 86

vasodilators: in Ménière's disease, 152, 154; in noise-induced hearing loss, 160; in sudden deafness, 164

ventilation tubes. See ear tubes

vertigo, 8, 9–10, 71–72, 220; acoustic neuroma and, 171, 172; airplane ear and, 203; autoimmune inner ear disease and, 165, 191; balance system disorders and, 71–72, 217, 220; barotrauma and, 195, 197; benign, of childhood, 78; benign paroxysmal positional, 10, 17, 71, 74–75, 188; brain disorders and, 78–79; causes, 10, 17; cholesteatoma and, 142, 217; drug-induced, 79, 168, 214; duration, 10; epidemic, 76; evaluation, 73; hypoglycemia and, 79; in labyrinthitis, 75–76; medications for, 78, 147, 151–52, 219; in Ménière's disease, 10, 17, 74, 145, 146, 147–48, 149, 218; migrainous, 77; perilymph fistula and, 77, 198;